Constellations Activity Book

by Ryan Jacobson
illustrations by Shane Nitzsche

Adventure Publications, Inc.
Cambridge, MN

Cover design by Jonathan Norberg
Edited by Dan Downing

10 9 8 7 6 5 4 3

Published by Adventure Publications, Inc.
820 Cleveland Street South
Cambridge, MN 55008
1-800-678-7006
www.adventurepublications.net

Printed in U.S.A.

ISBN: 978-1-59193-325-0

CONSTELLATIONS

What is a constellation?

People have gazed at stars for thousands of years. Some people used their imaginations and saw patterns and shapes in the stars. The patterns were given names and stories, and they became our constellations. A constellation is a group of stars that form a pattern or shape.

There are 88 constellations in all. Many of them are from thousands of years ago. Back then, people read the stars for two big reasons. Farmers used stars as a calendar, so they knew when it was time to plant their crops. Travelers used stars as a compass, so they always knew which direction they were going. The constellations and their stories helped people learn the stars and remember them.

Why do constellations move?

Constellations don't move. We do. The earth spins around once every day, and it orbits around the sun once every year. Imagine looking at something on a table. Now imagine spinning in circles while walking around the table. Your view of the object would change, right? This is also why our view of constellations changes.

It's important to remember: Constellations aren't in the same place every night. They appear to move, rotate and get higher or lower in the sky. Some get so low that we cannot see them at all.

How do I find each constellation?

In the bottom left corner, instructions are given for each constellation. These instructions are for about 10 p.m. each night. (If it is earlier in the evening, look lower and more to the east.) They tell you what direction to face, how high to look and what time of year to do it. In most cases, only two or three months are given. But most constellations can be seen several months of the year.

You can look for a constellation outside the noted months. Here are a few tips*:

1. The constellation will be lower in the night sky.
2. In the months before those given, look a little more to the east.
3. In the months after those given, look a little more to the west.
4. If you cannot find it, remember that it may not be visible at that time of night or year.

*These tips do not always work for the four constellations in this book that can be seen every night of the year: Cassiopeia (page 42), Draco (page 20), Ursa Major (page 10) and Ursa Minor (page 6).

Where did the stories come from?

Forty-eight of the constellations were described by a Greek astronomer, Ptolemy. He wrote a book about them nearly 2,000 years ago. Most of the stories we know come from him and his time. Many of them are based on his people's religions and beliefs. We call those stories Greek mythology.

What is Greek mythology?

Thousands of years ago, little was known about science and nature. The people of Greece did not understand things like the weather. They explained them by telling stories about powerful gods. The stories became a religion. According to Greek mythology, there were many different gods and goddesses. Each was in charge of something different. The twelve main gods and goddesses lived on top of Mount Olympus. Zeus was the leader of the gods. His wife, Hera, was the goddess of marriage. Hades was Zeus's brother, and he ruled the Underworld—the land of the dead. And so on. Greek mythology was filled with heroes and adventures. The gods and goddesses played a part in most stories. Sometimes they did very good things. Sometimes they did very bad things.

Are the stories true?

No, they are just stories.

What are the zodiac constellations?

There are 12 constellations of the zodiac, and they are special because of their location. The earth orbits around the sun at a certain level, or plane, in space. The zodiac constellations are the constellations at that same level.

Here's another way to think about it. The zodiac constellations form a giant circle around the sun and earth. Inside that circle, the earth moves around the sun—while the sun and the constellations do not appear to move. This means that the sun is always between the earth and one of the zodiac constellations. So when the sun rises each morning, it seems to rise into a zodiac constellation. Some people believe we each have a zodiac constellation, or sign. Your sign would be the zodiac constellation that the sun rose into on the day you were born. (See *What's Your Sign?*, page 9.)

Is every constellation in this book?

There are 88 constellations in all. This book includes 26 of them. The ones chosen here include a mix of the most famous constellations, the easiest to find and the most important to know.

CONSTELLATIONS MAP

The map at right shows the 26 constellations in this book. Since the page is flat and the night sky isn't, the map is not exact. But the map can still be helpful. It gives an idea of each constellation's size compared to the other constellations. It also shows where each constellation is in relation to the other constellations.

Remember that you will never be able to see all of the constellations at the same time. In fact, different constellations are best seen at very different times of the year. (Plus, sometimes only part of a constellation is visible.) The chart below tells you the months that you will be able to see each of these 26 constellations:

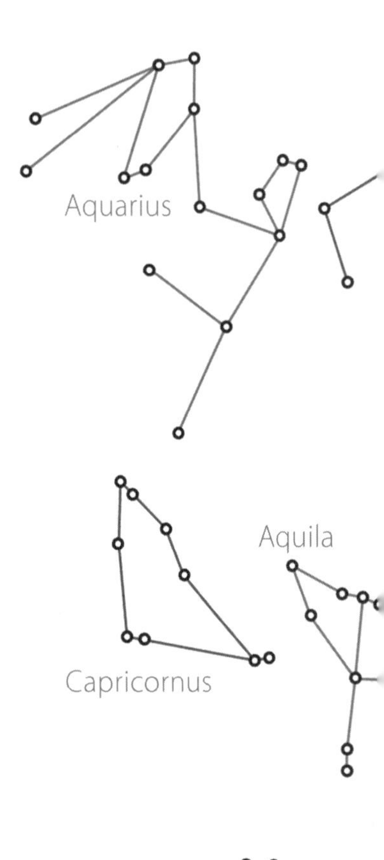

CONSTELLATION	MONTHS VISIBLE	GENERAL DIRECTIONS	PAGE
Andromeda	June to March	NE, N, NW	38
Aquarius	June to December	SE, S, SW	34
Aquila	May to November	E, S, W	28
Aries	August to March	E, S, W	44
Boötes	February to September	E, S, W	16
Cancer	November to June	E, S, W	56
Canis Major	November to April	SE, S, SW	52
Capricornus	July to November	SE, S, SW	32
Cassiopeia	Every month	NE, N, NW	42
Cygnus	April to December	NE, N, NW	30
Draco	Every month	NE, N, NW	20
Gemini	October to May	E, S, W	54
Hercules	March to October	E, S, W	24
Hydra	January to June	SE, S, SW	12
Leo	December to July	E, S, W	8
Libra	April to August	SE, S, SW	18
Orion	October to April	SE, S, SW	50
Pegasus	June to January	E, S, W	36
Perseus	July to April	NE, N, NW	46
Pisces	July to January	E, S, W	40
Sagittarius	June to September	S	26
Scorpius	May to August	S	22
Taurus	September to April	E, S, W	48
Ursa Major	Every month	NE, N, NW	10
Ursa Minor	Every month	N	6
Virgo	February to July	E, S, W	14

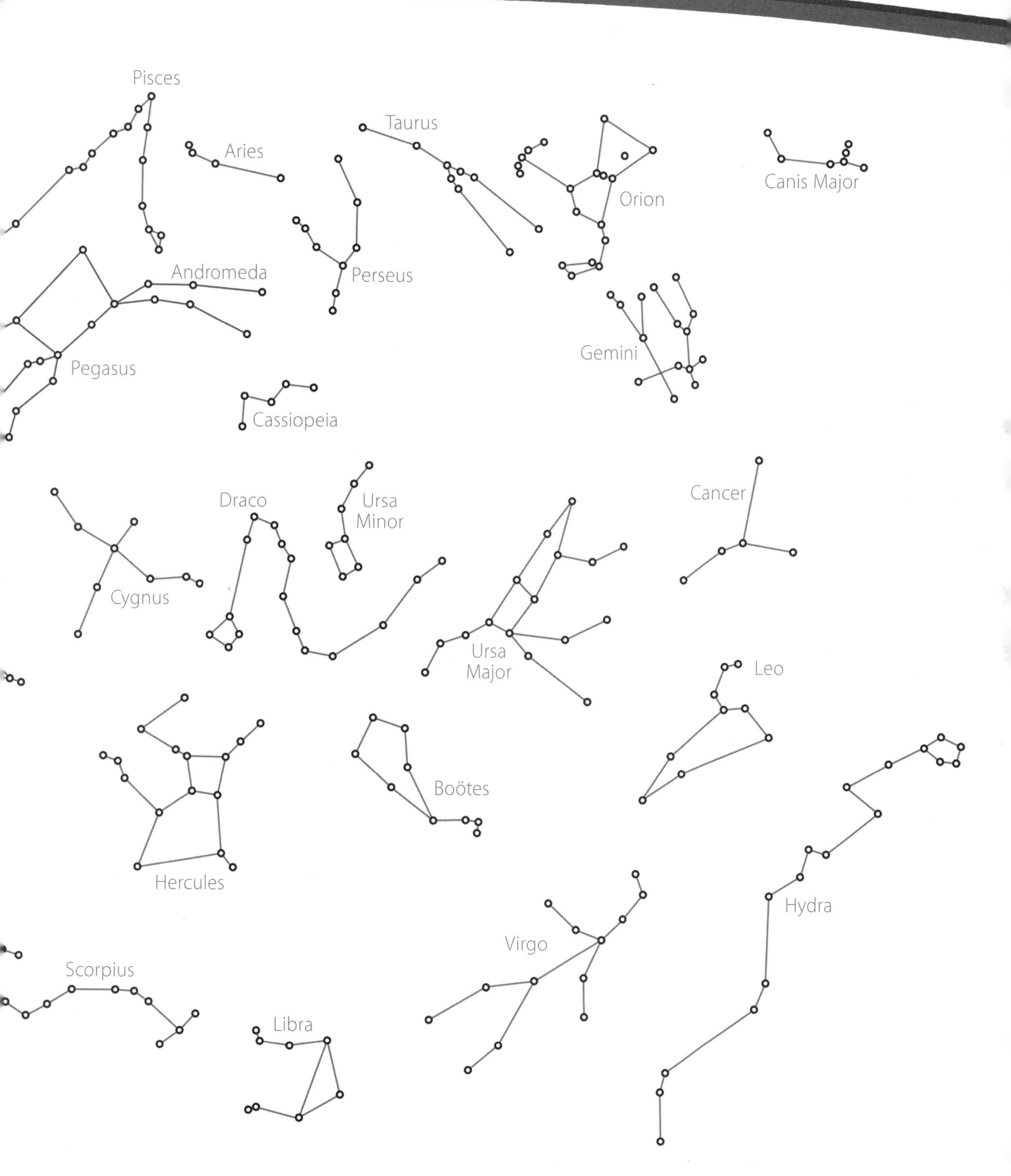
Pisces
Aries
Taurus
Orion
Canis Major
Perseus
Andromeda
Gemini
Pegasus
Cassiopeia
Cancer
Draco
Ursa Minor
Cygnus
Ursa Major
Leo
Boötes
Hercules
Hydra
Virgo
Scorpius
Libra

URSA MINOR

THE LITTLE BEAR

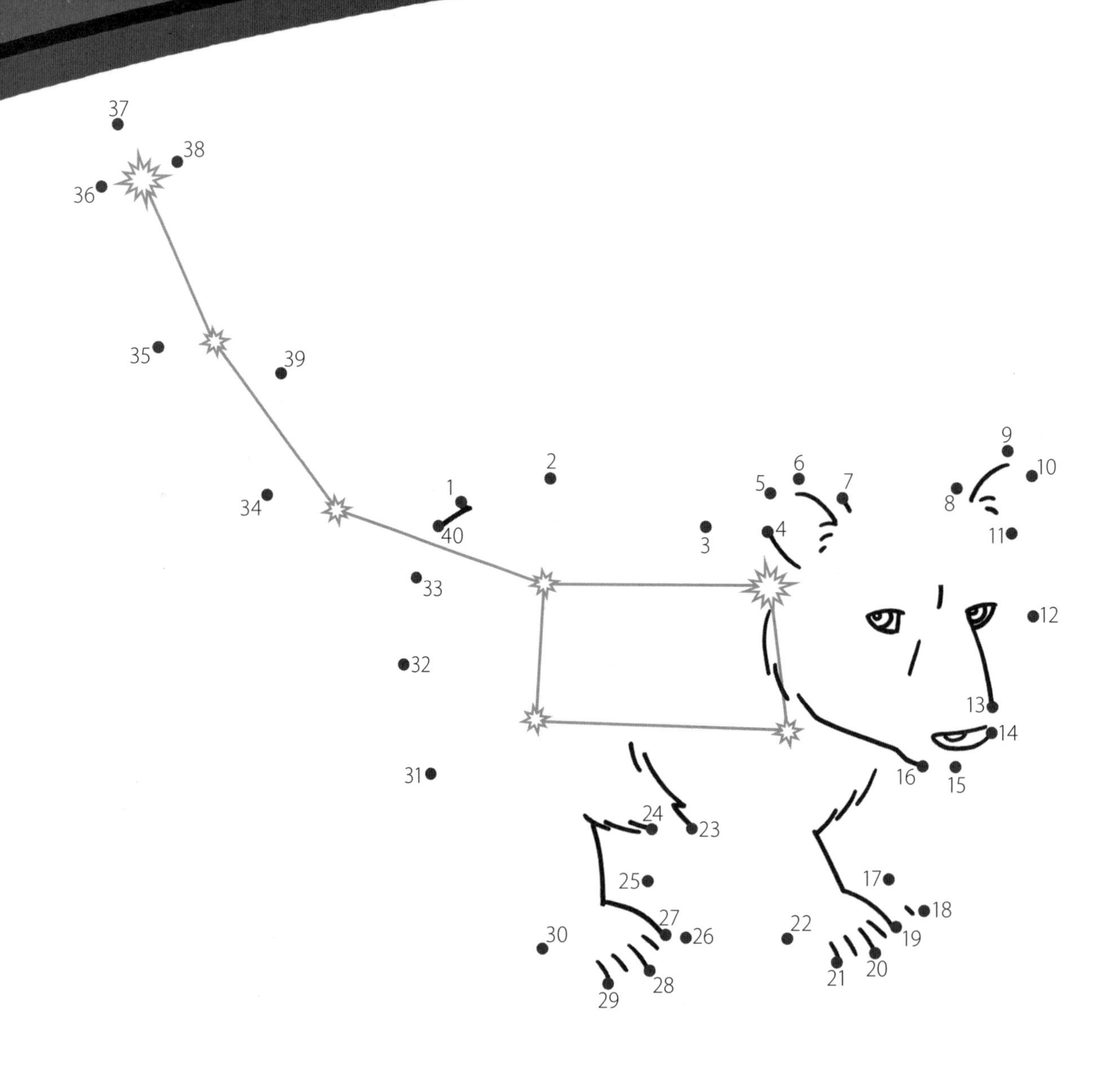

Best Way to Find It (about 10 p.m.)

Look this high . . . any night of the year . . . facing north . . . for this star pattern.

TIP: Every night, you will find Ursa Minor near the same spot, but it may be sideways or upside down.

Ursa Minor is home to the Little Dipper. The North Star is at its tip.

As the Story Goes

Arcas was a great hunter. His mother, Callisto, was turned into a bear, and Arcas accidentally hunted her. The god Zeus saved Callisto by pulling her into the sky. Arcas was also changed into a bear, Ursa Minor (*ER-suh MY-nor*). The gods lifted him by his tail, stretching it. They put him in the stars to be with his mother.

ALSO CHECK OUT: Ursa Major (page 10)

Name the Constellations

Ursa Minor is a well-known constellation. Do you recognize any of these others? When you find them in the activity book, come back and write their names on the blanks below.

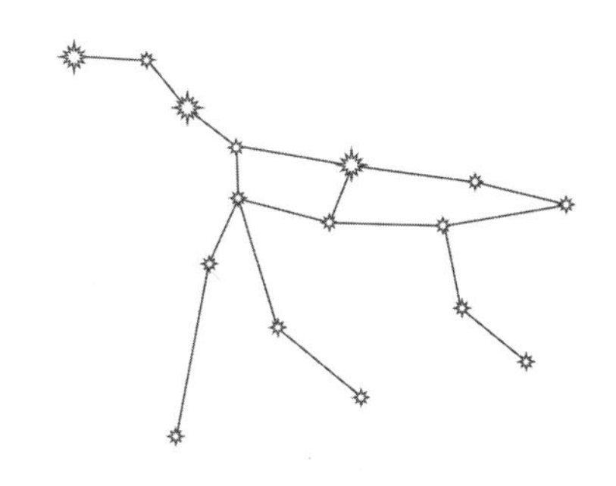

1. ____________________

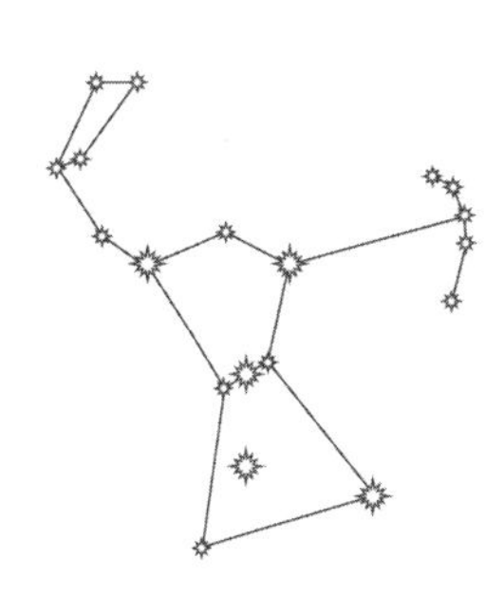

2. ____________________

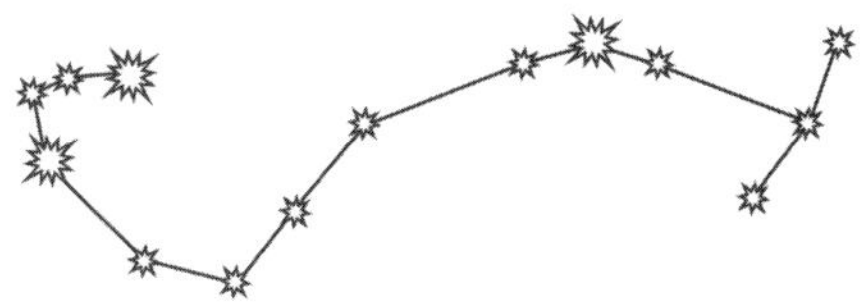

3. ____________________

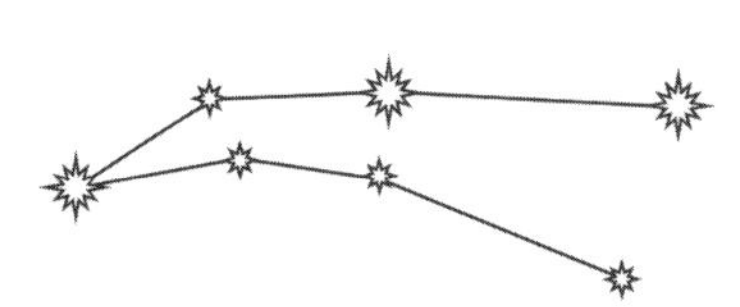

4. ____________________

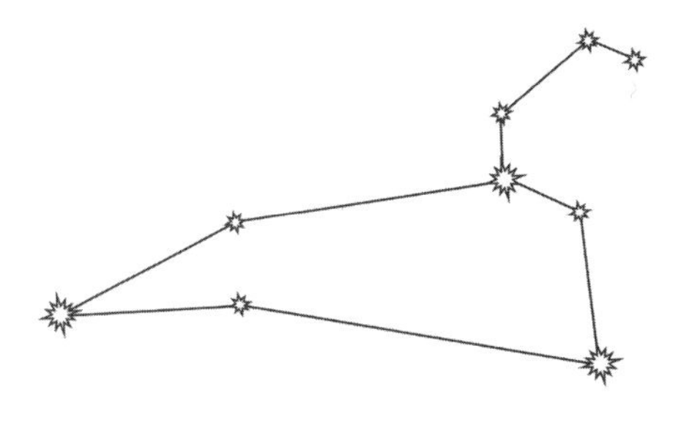

5. ____________________

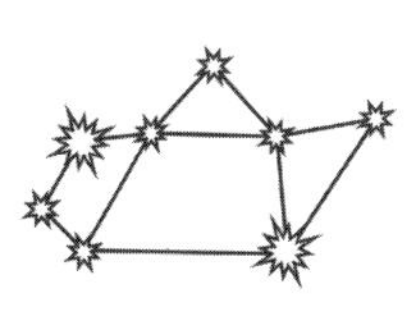

6. ____________________

(answers on page 58)

LEO

THE LION

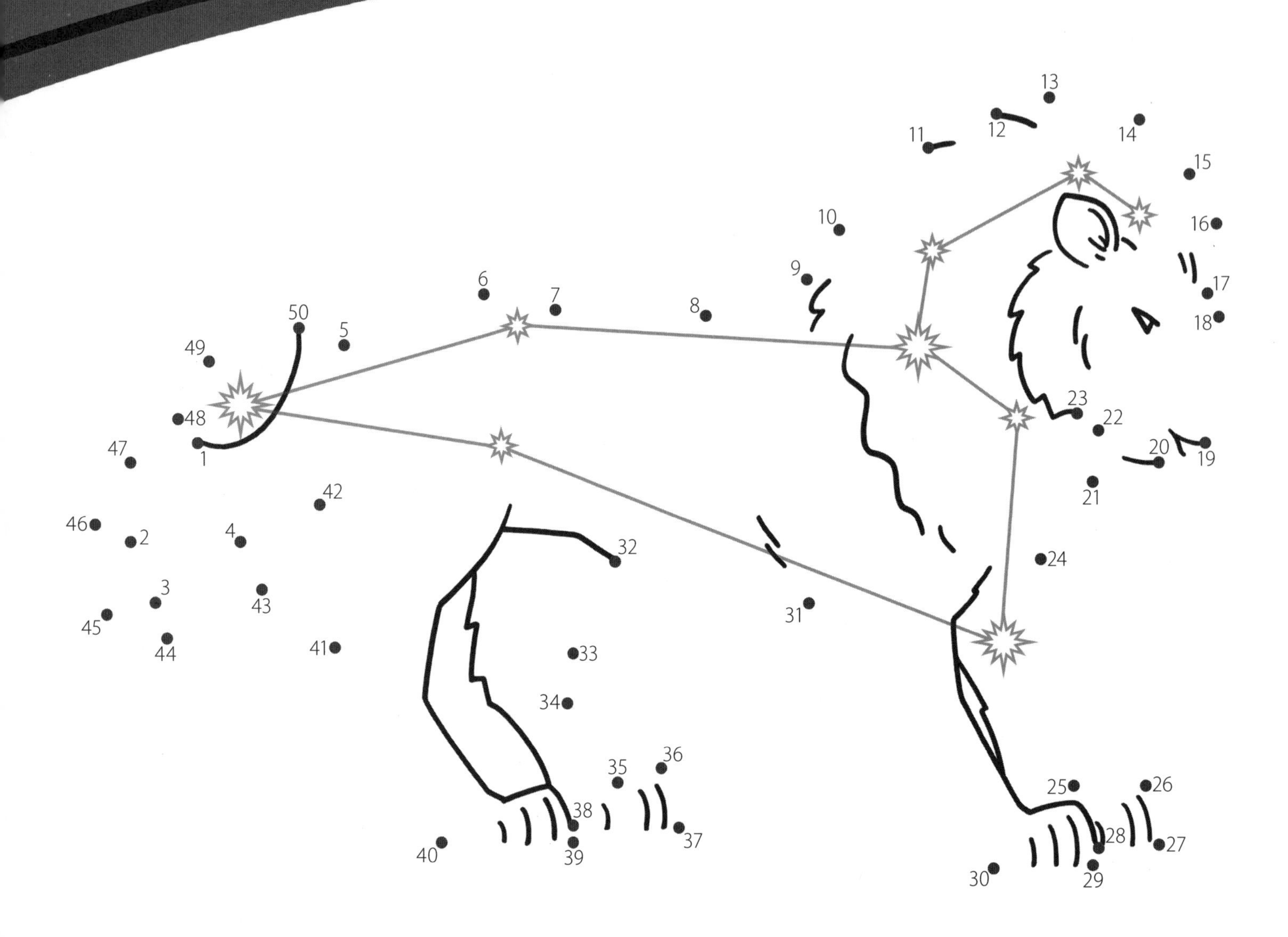

Best Way to Find It (about 10 p.m.)

Look this high . . . March to April . . . facing south . . . for this star pattern.

TIP: Ursa Major's straight back leg points right toward Leo.

Simple to spot, the constellation actually forms the shape of a lion.

As the Story Goes

Leo (*LEE-oh*) was no ordinary lion. It was a monster that fell to the earth from the moon. Its skin was stronger than armor, and no sword could hurt it. The hero Hercules was forced to battle Leo. Hercules wrestled the lion and defeated it, using the lion's own claws to cut its skin.

ALSO CHECK OUT: Cancer (page 56), Draco (page 20), Hercules (page 24), Hydra (page 12)

What's Your Sign?

Some people think your birth date tells a lot about you. For this reason, there is a constellation—called the zodiac sign—for every date. Just for fun, find your sign below. Does it describe you?

BIRTH DATES	CONSTELLATION	DESCRIPTION
March 21 to April 19	**Aries** (page 44)	Brave, leader, always in a hurry, likes to compete, likes to work alone
April 20 to May 20	**Taurus** (page 48)	Kind, creative, strong-minded, usually very happy, likes to have fun
May 21 to June 20	**Gemini** (page 54)	Curious, clever, picky, likes to be with others, likes things to change
June 21 to July 22	**Cancer** (page 56)	Caring, loyal, shy, likes to take care of others, likes to be with family
July 23 to August 22	**Leo** (page 8)	Confident, creative, strong-minded, likes to have fun, likes to be rewarded
August 23 to September 22	**Virgo** (page 14)	Proud, loyal, timid, likes to help others, always wants to get better at things
September 23 to October 22	**Libra** (page 18)	Honest, friendly, likes to have things his or her way, likes music, likes to laugh
October 23 to November 21	**Scorpio** (page 22)	Smart, brave, sneaky, likes to have secrets, wants to succeed
November 22 to December 21	**Sagittarius** (page 26)	Happy, proud, doesn't like schedules, deals well with change, likes to travel
December 22 to January 19	**Capricorn** (page 32)	Smart, patient, doesn't like things to change, serious, hard-working
January 20 to February 18	**Aquarius** (page 34)	Creative ideas, leader, wild, works well with others, likes to get things done
February 19 to March 20	**Pisces** (page 40)	Caring, creative, shy, emotional, likes to make others happy

URSA MAJOR

THE GREAT BEAR

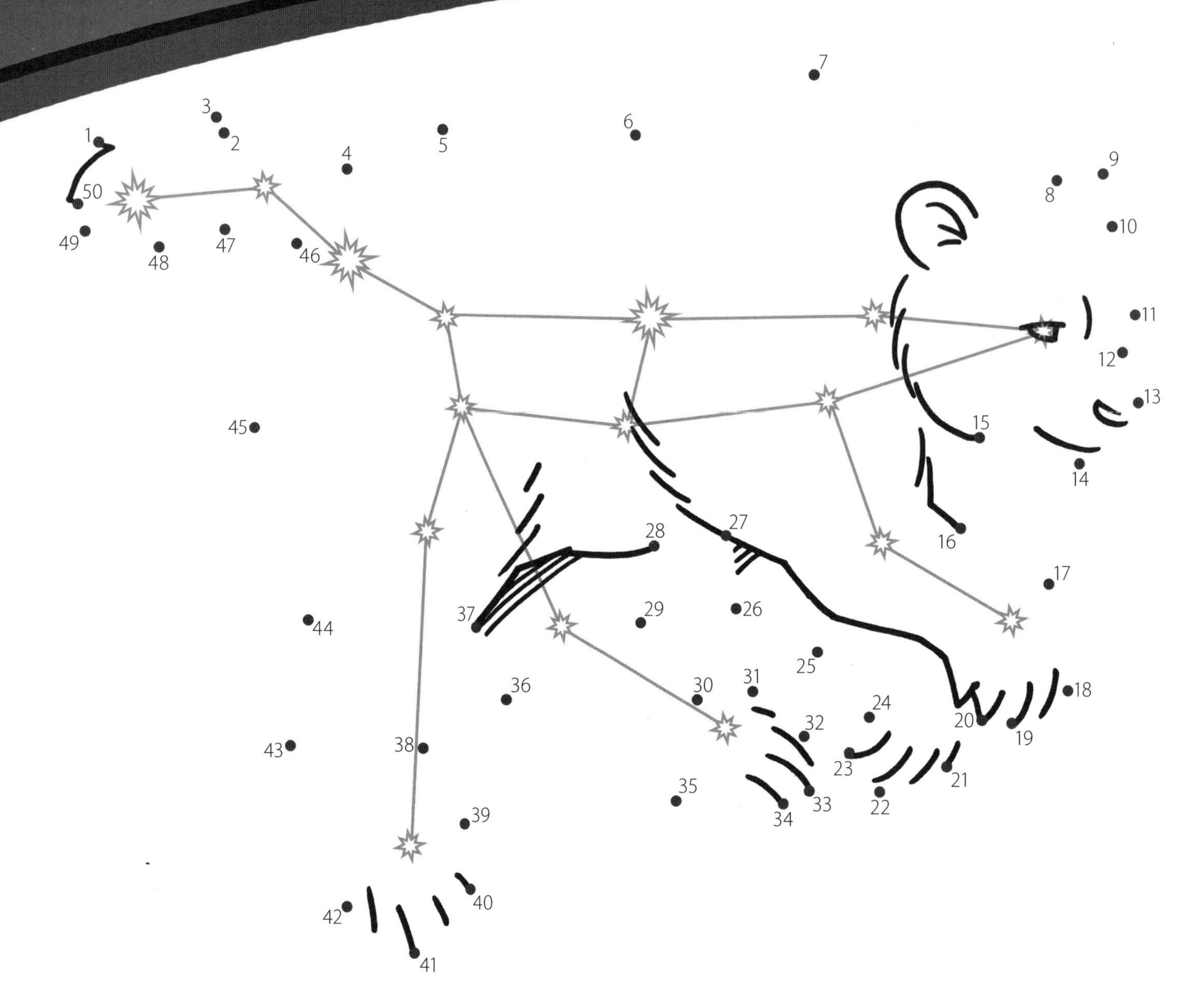

Best Way to Find It (about 10 p.m.)

Look this high . . . March to May . . . facing north . . . for this star pattern.

TIP: In April, Ursa Major is almost straight above you. By October, it's barely above the horizon.

The Big Dipper is part of Ursa Major. It makes up her tail and back end.

As the Story Goes

Callisto fell in love with the god Zeus, but Zeus was married to a goddess named Hera. Hera got mad at Callisto and turned her into a bear, Ursa Major (*ER-suh MAY-jur*). A hunter named Arcas chased Ursa Major into a sacred temple, but Zeus saved the bear. He pulled Ursa Major into the sky by her tail. That is why she has a long tail.

ALSO CHECK OUT: Ursa Minor (page 6)

Escape from Arcas

Help Ursa Major find her way through the maze to the sacred temple.

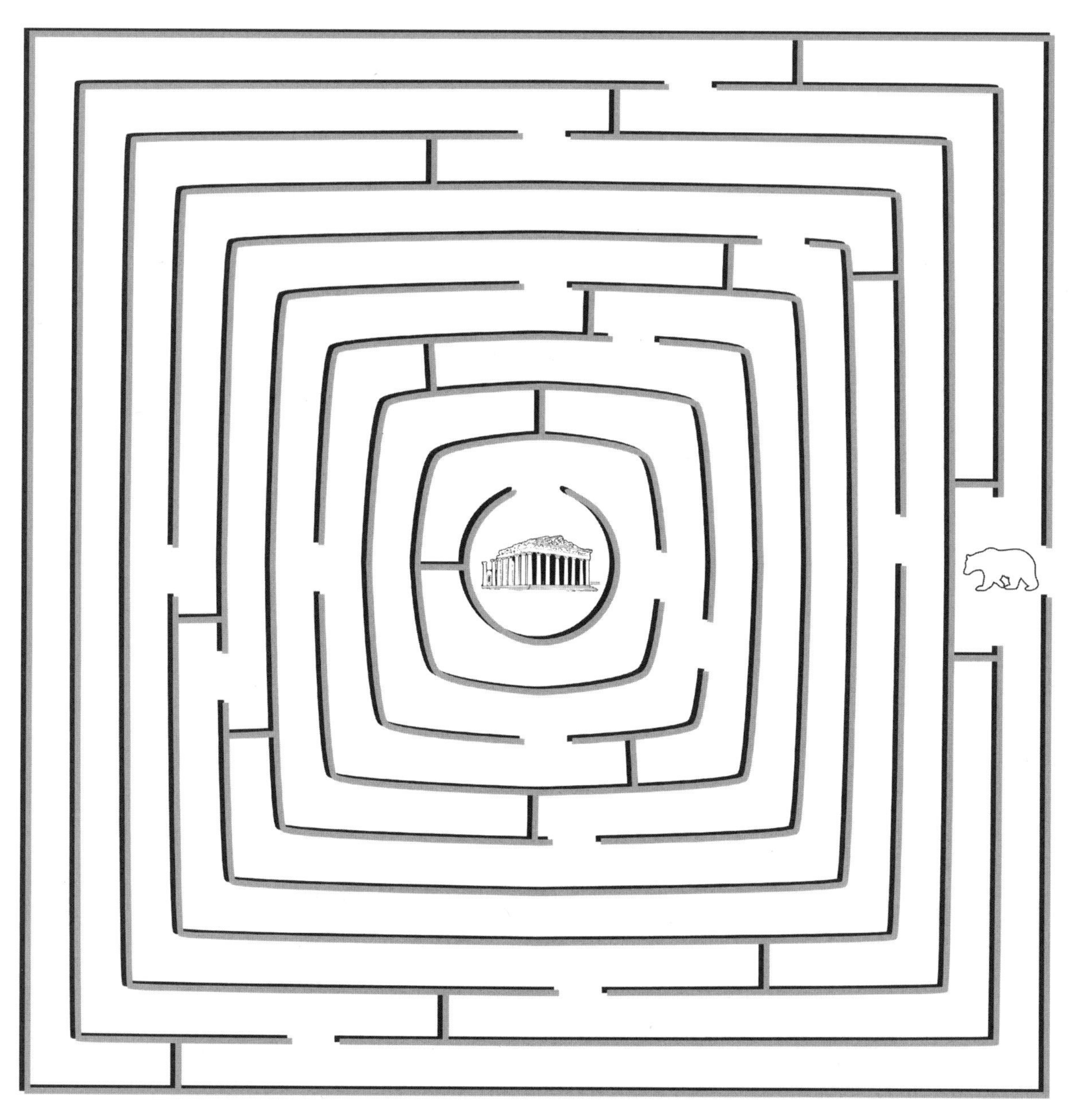

(answers on page 58)

HYDRA

THE SERPENT

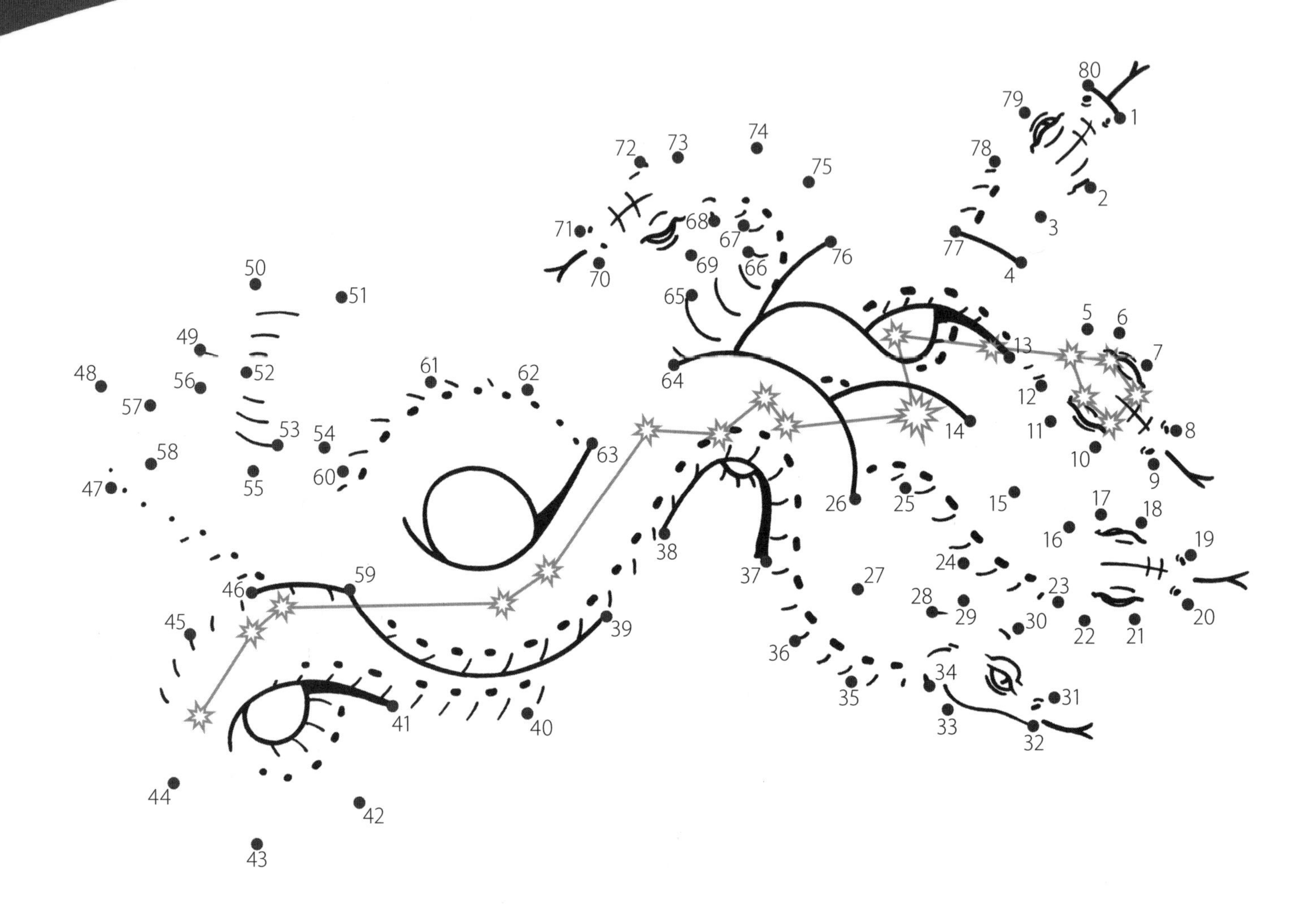

Best Way to Find It (about 10 p.m.)

Look this high . . . March to May . . . facing south . . . for this star pattern.

TIP: Hydra is the biggest constellation. You may not be able to see it all at once.

Hydra stretches far across the night sky, from east to west.

As the Story Goes

Hydra (*HIGH-druh*) was a terrible monster. It had many heads and poison breath. The hero Hercules battled Hydra. Hercules covered his mouth as he chopped off the serpent's heads. But every time Hydra lost a head, two more grew in its place. Instead of a sword, Hercules used a torch to finally defeat the monster.

ALSO CHECK OUT: Cancer (page 56), Draco (page 20), Hercules (page 24), Leo (page 8)

Don't Lose Your Head

Hydra has many heads. On the picture below, draw more heads to match the first one. And remember that you can draw more than one head at each neck.

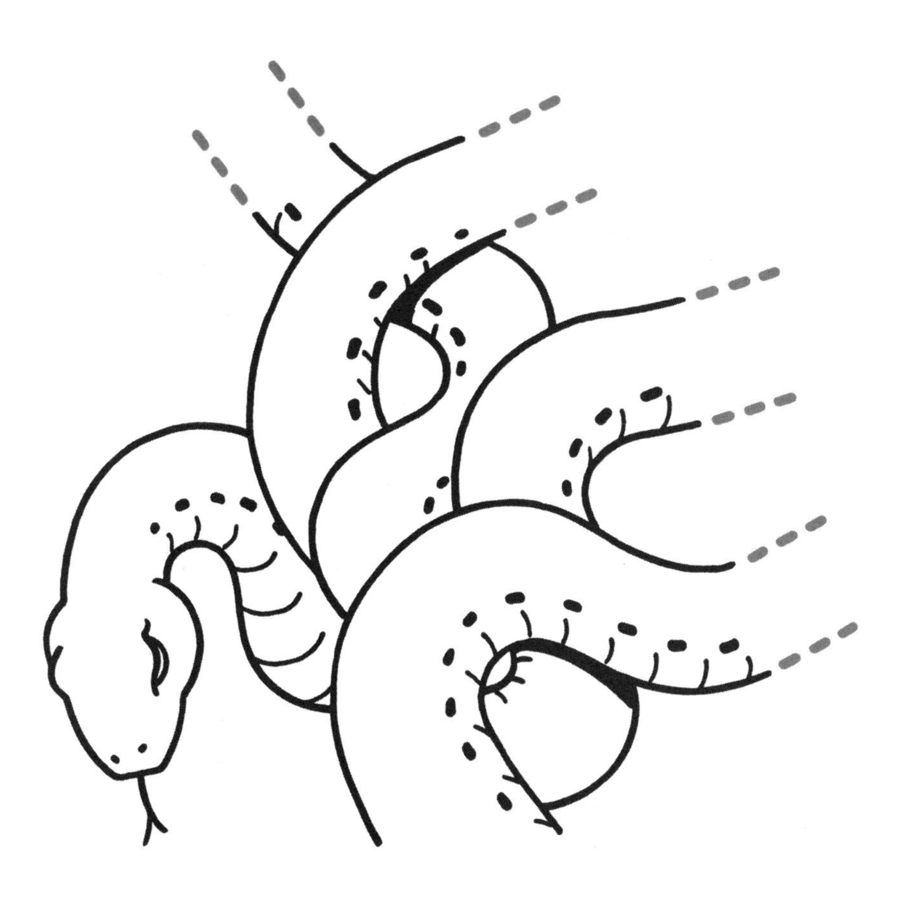

VIRGO

THE MAIDEN

Best Way to Find It (about 10 p.m.)

Look this high . . . April to June . . . facing south . . . for this star pattern.

TIP: Virgo is the second-biggest constellation. Its stars are very spread out.

Virgo's stars make the shape of a giant stick figure.

As the Story Goes

Virgo (*VUR-go*) was the goddess of justice, known as Dikē. She ruled as a queen on earth, and it was a time of great peace and joy. But the humans became greedy. They were no longer happy, and they began fighting wars. Dikē grew sad. She left earth to live with her father, the god Zeus.

ALSO CHECK OUT: What's Your Sign? (page 9)

Many Maidens

In our story, Virgo was the goddess Dikē. But the constellation has stood for many different maidens in many different parts of the world. In the word find below, circle these names for Virgo.

A T A R G A T I S H A A R Y
J T S Y A T H T M E C E P I
P U H E T A J I Y A E M J S
E V S E N A D S T A R R U P
R D E I N E I O R T E Y H R
S R O S P A K H E C S I M O
E I Y H E S E R C E A A T S
P S U O R T T A T H P R O E
H U H L S R I D E M E T E R
O S I A T A A I S P P R I P
N O S U L E D E T R E L A I
E T I A H A R T H E M I S N
T I S R E S U M I U S T I A
D F O R T U N A T A M A A T

ASTRAEA	DIKE	MARY
ATARGATIS	FORTUNA	PERSEPHONE
ATHENA	HERA	PROSERPINA
CERES	ISIS	SHALA
DEMETER	MAAT	THEMIS

(answers on page 58)

BOÖTES

THE HERDSMAN

Best Way to Find It (about 10 p.m.)

Look this high . . . | May to June . . . | facing south . . . | for this star pattern.

TIP: Ursa Major's tail always points toward Boötes.

Boötes looks like a huge kite, blowing in the wind.

As the Story Goes

Demeter was the goddess of agriculture, or farming. Boötes (*boe-OH-teez*) was her son. He saw how hard it was to dig up the land for planting seeds, so he made a special invention. Farmers could push his invention across the ground and slice a deep path in the soil. This made planting much easier.

What Did Boötes Invent?

Use the pictures as clues to fill in the blanks below. Then write the letters that match the numbers at the bottom of the page to find out what Boötes invented.

_ _   _ _   V   _ _ _ _ _   _ _ _   _ _ _ _ .

1 2 3 4   2 4 5 2 6   5 1 2   7 8 9 10

(answers on page 58)

LIBRA

THE BALANCE

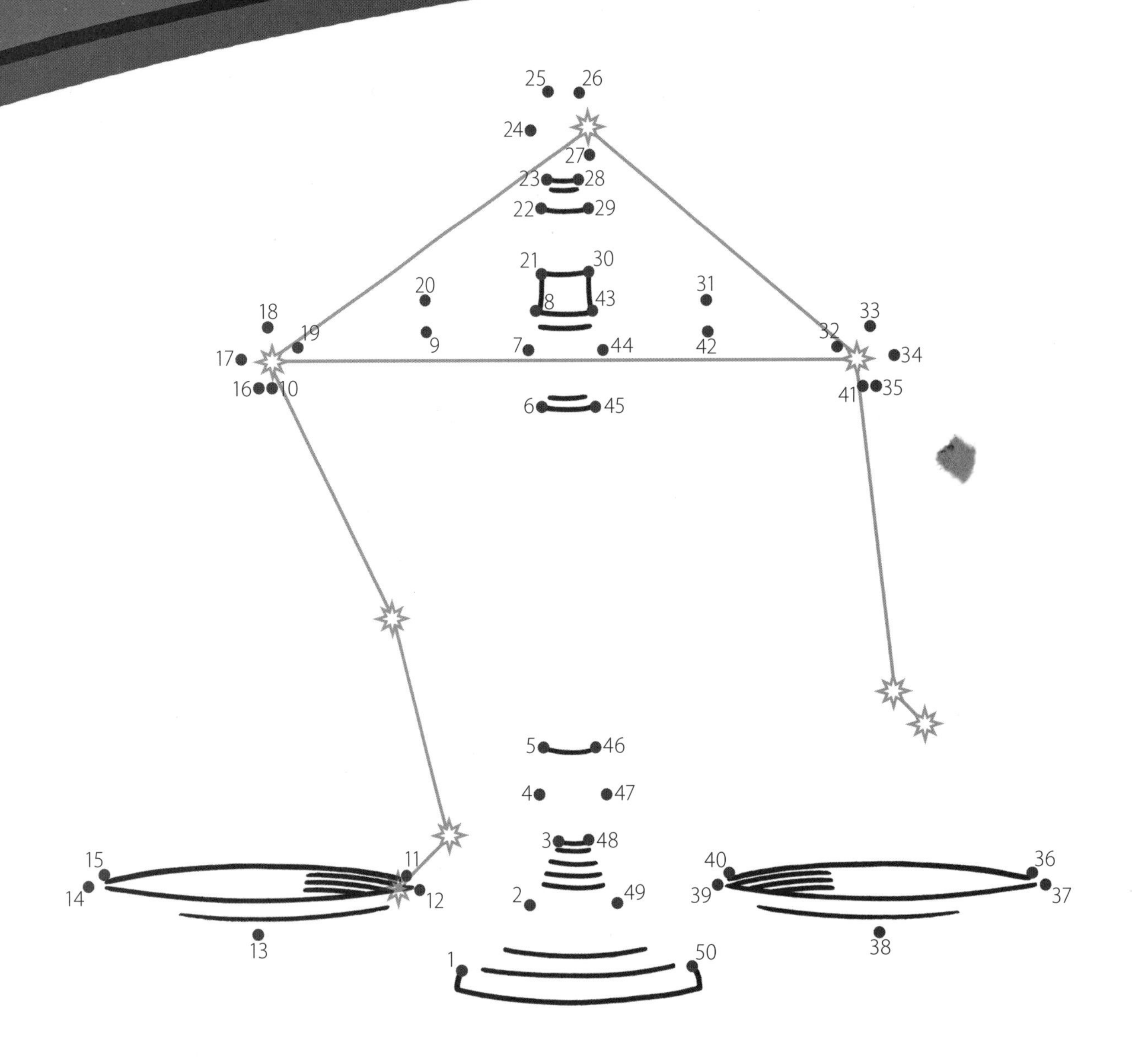

Best Way to Find It (about 10 p.m.)

Look this high . . . May to July . . . facing south . . . for this star pattern.

TIP: Libra's stars are not very bright. Look for them near Virgo's feet.

As the Story Goes

Libra (*LEE-bruh*) is a scale, or a balance. It stands for fairness and justice. Thousands of years ago, Libra's stars were said to form a different constellation: Hades' Golden Chariot. Hades was the Lord of the Underworld. He used his chariot to travel to earth. During one visit, he kidnapped Zeus's daughter Persephone, causing winter to come.

ALSO CHECK OUT: What's Your Sign? (page 9)

Which Weighs More?

The items below would not balance Libra's scale. Circle the object in each pair that weighs more.

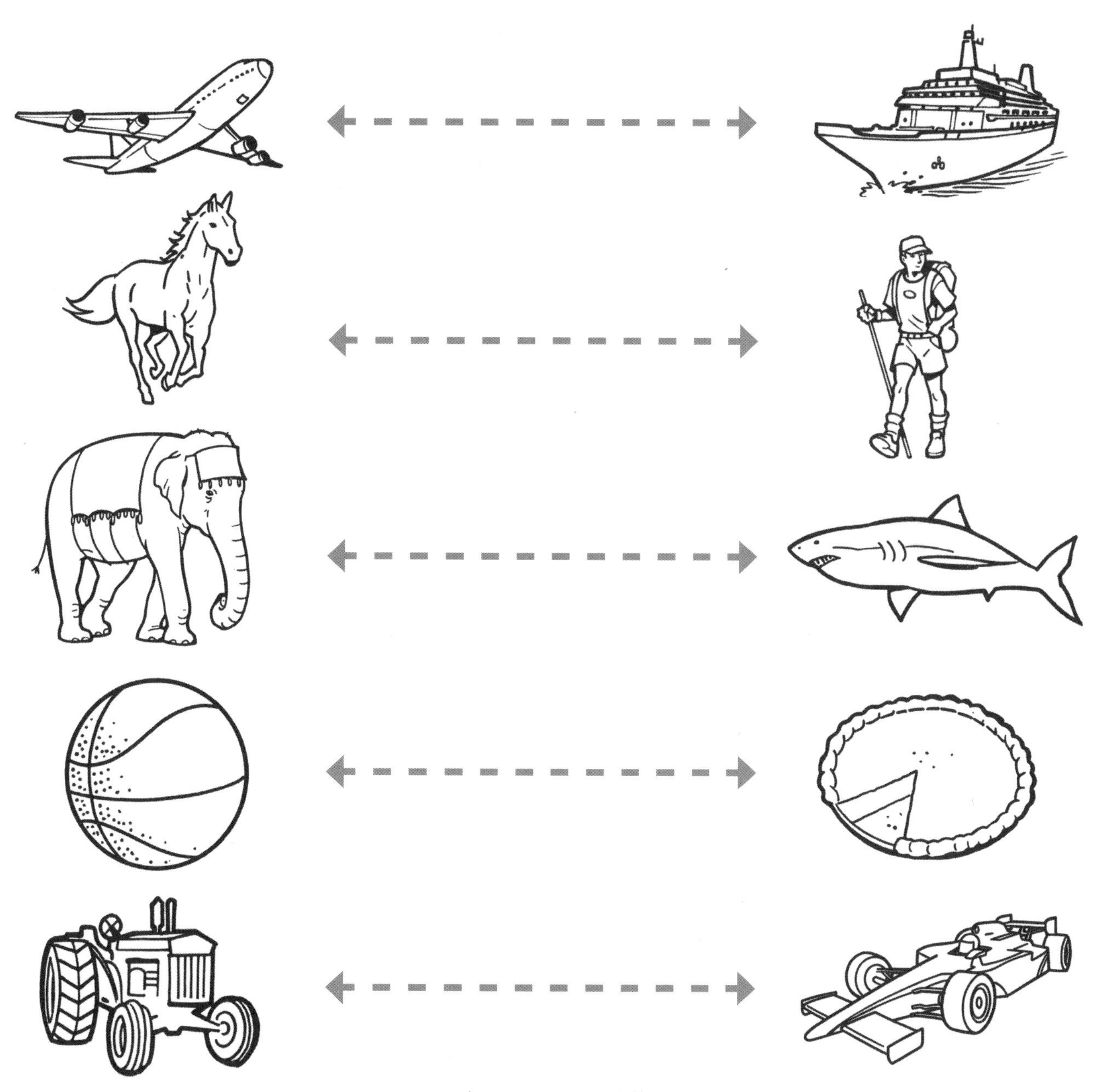

(answers on page 58)

DRACO

THE DRAGON

Best Way to Find It (about 10 p.m.)

Look this high . . . May to July . . . facing north . . . for this star pattern.

TIP: Draco's tail lies between Ursa Major and Ursa Minor.

Draco is quite large, but its stars do not shine very brightly.

As the Story Goes

The hero Hercules was a slave. In order to be free, he had to do twelve jobs, or labors. One was to pick golden apples from a magic tree. But Draco (*DRAY-co*), a dragon with 100 heads, guarded the apples. Hercules defeated Draco by shooting it with a poison arrow. He brought the golden apples to his master and gained his freedom.

ALSO CHECK OUT: Cancer (page 56), Hercules (page 24), Hydra (page 12), Leo (page 8)

Constellation Bingo

Mark each constellation with an X as you find it in a field guide or in the night sky. (It may take weeks or even months to find them all in the night sky.) When you get five in a row, you win!

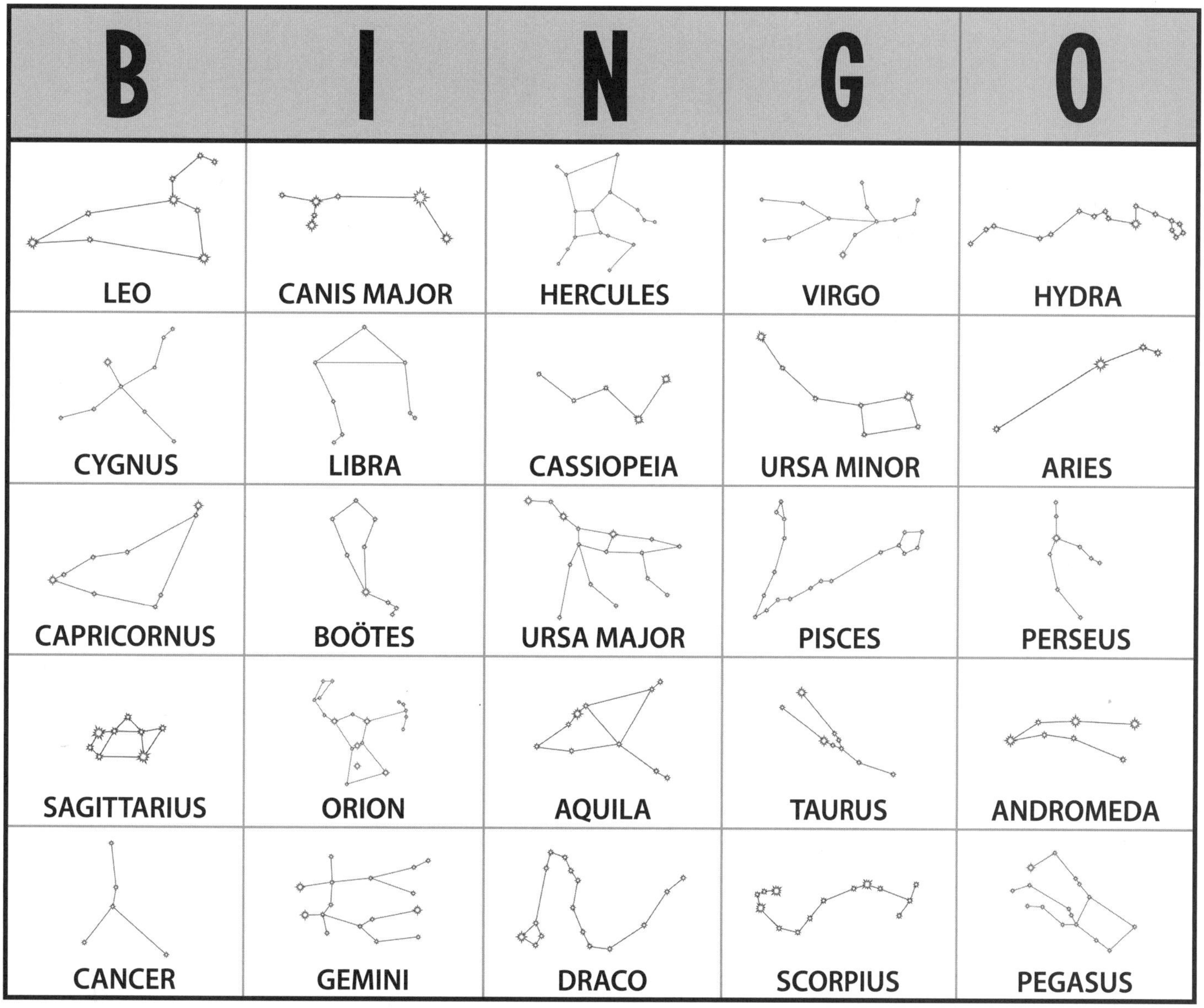

B	I	N	G	O
LEO	CANIS MAJOR	HERCULES	VIRGO	HYDRA
CYGNUS	LIBRA	CASSIOPEIA	URSA MINOR	ARIES
CAPRICORNUS	BOÖTES	URSA MAJOR	PISCES	PERSEUS
SAGITTARIUS	ORION	AQUILA	TAURUS	ANDROMEDA
CANCER	GEMINI	DRACO	SCORPIUS	PEGASUS

SCORPIUS

THE SCORPION

Best Way to Find It (about 10 p.m.)

Look this high . . . May to August . . . facing south . . . for this star pattern.

TIP: Look for the bright star in the middle of Scorpius's body. It shines with a red tint.

As the Story Goes

Orion, the great hunter, bragged that he could defeat any animal. So the goddess Hera decided to test him. She sent a tiny scorpion to challenge the hunter. Scorpius (*SCOR-pee-us*) stung Orion. Before Orion died, he smashed Scorpius with his club. The gods placed Scorpius in the sky to honor his bravery.

ALSO CHECK OUT: Orion (page 50), What's Your Sign? (page 9)

Amazing Animal Tails

Use the clues below to solve the crossword puzzle about animals with interesting tails.

ACROSS

1. A tail, or a toy for a baby?
6. This cat may erase its tracks with its spotted tail
7. It hangs upside down from its furless tail
10. This animal hits fish out of the water with its tail
11. Rodents whose tails keep their bodies the right temp
12. Its tail helps it grip and balance as it climbs trees
13. Beware this animal's tail, or you could get sprayed

DOWN

1. Its tail is bushy with black and gray rings
2. Kingly beasts whose tails tell how they're feeling
3. When this animal is in danger, its tail falls off
4. It's not a beaver, but it has a tail like one
5. This big reptile uses its tail as a dangerous weapon
8. Scorpius, for one
9. This meat-eater's tail helps it swim really fast

(answers on page 58)

HERCULES

THE HERO

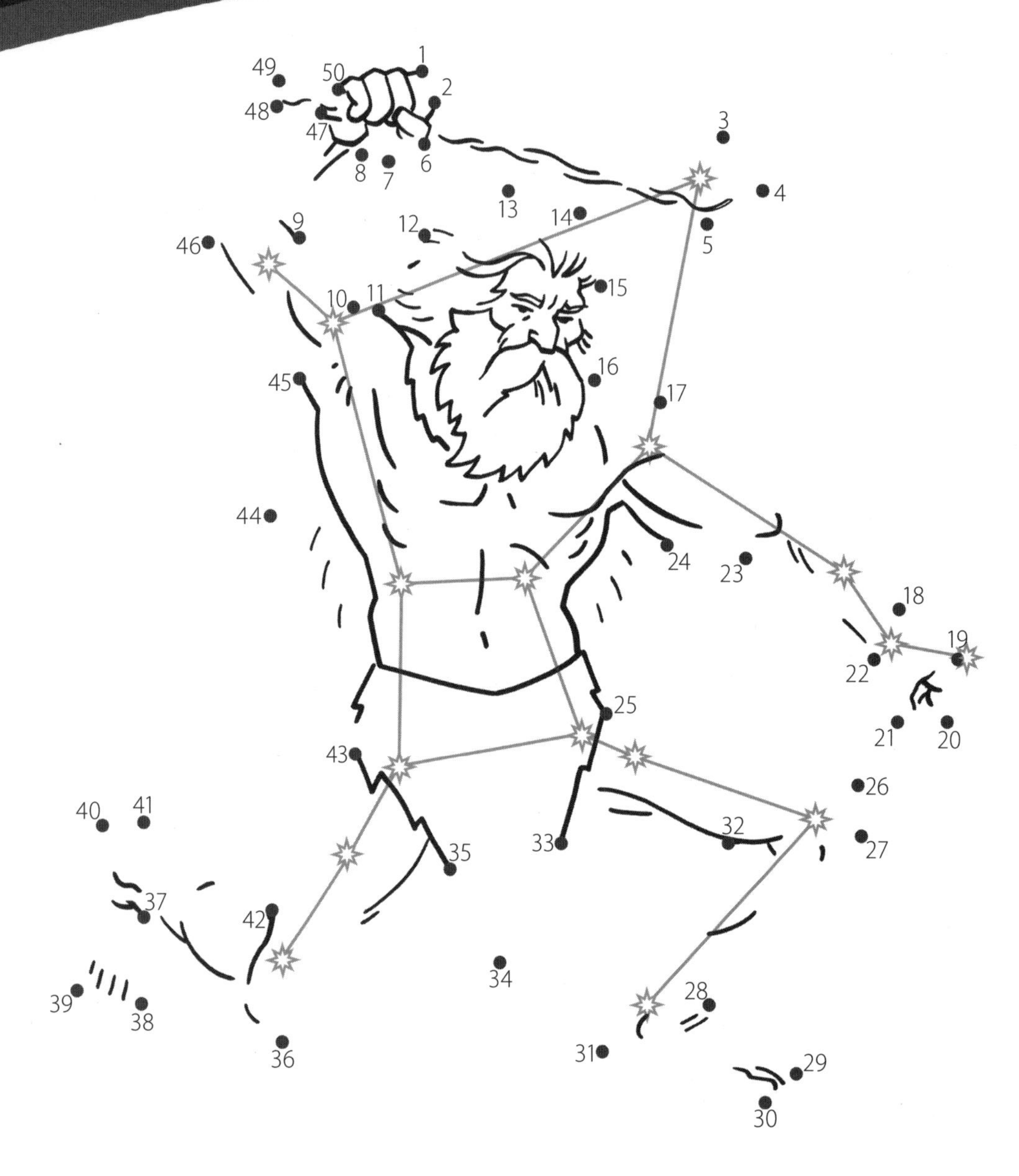

Best Way to Find It (about 10 p.m.)

Look this high . . . June to July . . . facing south . . . for this star pattern.

TIP: The constellation Boötes looks like a kite that is blowing toward Hercules.

Hercules' stars form the shape of a man. In the night sky, they are upside down.

As the Story Goes

Hercules (*HER-kyew-leez*) was the son of the god Zeus, and he was the strongest man alive. He didn't get along with his mother, Hera. Eventually, Hercules became a king's slave. He had to complete twelve hard tasks, or labors, to earn his freedom. Hera tried to stop him, but Hercules finished all twelve. He went on to have many adventures.

ALSO CHECK OUT: Cancer (page 56), Draco (page 20), Hydra (page 12), Leo (page 8)

The Twelve Labors of Hercules

To earn his freedom, Hercules completed twelve amazing tasks. Unscramble the words below, and find out what the tasks were.

1. Defeat a magical _ _ _ _ (I O L N), whose skin could not be cut.
2. Defeat the _ _ _ _ _ (R Y H A D), a monster with many heads.
3. Hunt for and catch the goddess Diana's favorite _ _ _ _ (E R E D).
4. Hunt for and catch a wild and mighty _ _ _ (I G P).
5. Clean the stables of a rich king's _ _ _ _ _ _ (T E T L A C) in one day.
6. Scare away a giant flock of _ _ _ _ _ (D R I S B).
7. Hunt for and catch an angry and savage _ _ _ _ (L U L B).
8. Capture another king's herd of _ _ _ _ _ _ (S H E O R S).
9. Steal a _ _ _ _ (T E L B) from the Amazons, women warriors.
10. Steal cattle from a three-headed _ _ _ _ _ _ _ (S O R M E N T).
11. Pick golden _ _ _ _ _ _ (P A S L E P) from a dragon's tree.
12. Capture Hades' pet, a three-headed _ _ _ (O G D).

(answers on page 58)

SAGITTARIUS

THE ARCHER

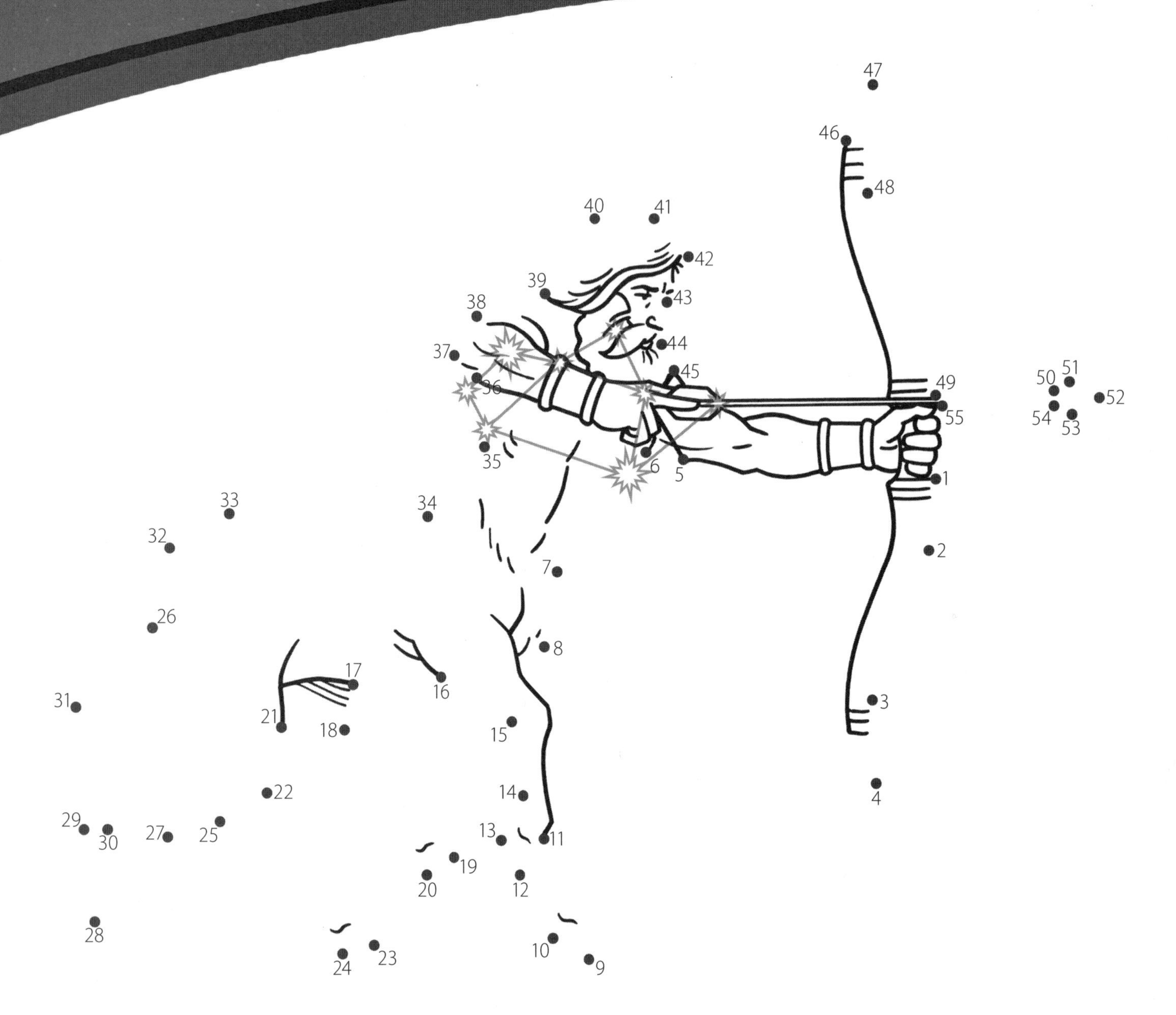

Best Way to Find It (about 10 p.m.)

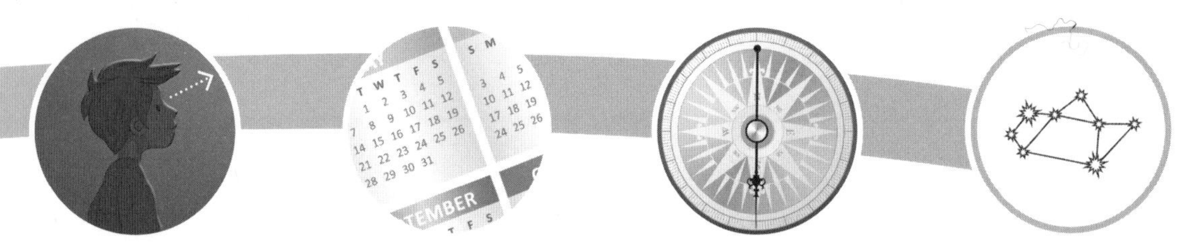

Look this high . . . June to September . . . facing south . . . for this star pattern.

TIP: You will always find Sagittarius next to the hook in Scorpius's tail.

As the Story Goes

Sagittarius (*SAJ-it-TARE-ee-us*), was a centaur—half man, half horse—and he was a son of the god Pan. He lived with nine magical women, called the Muses. They inspired him to become a great hunter, artist and inventor. As the story goes, he invented the bow and arrow. He also invented the beat of music by clapping to the Muses' songs.

ALSO CHECK OUT: What's Your Sign? (page 9)

Half and Half

A centaur was half person, half horse. A satyr was half person, half goat. Make up your own half-animals by drawing different animal bodies on Sagittarius below.

AQUILA

THE EAGLE

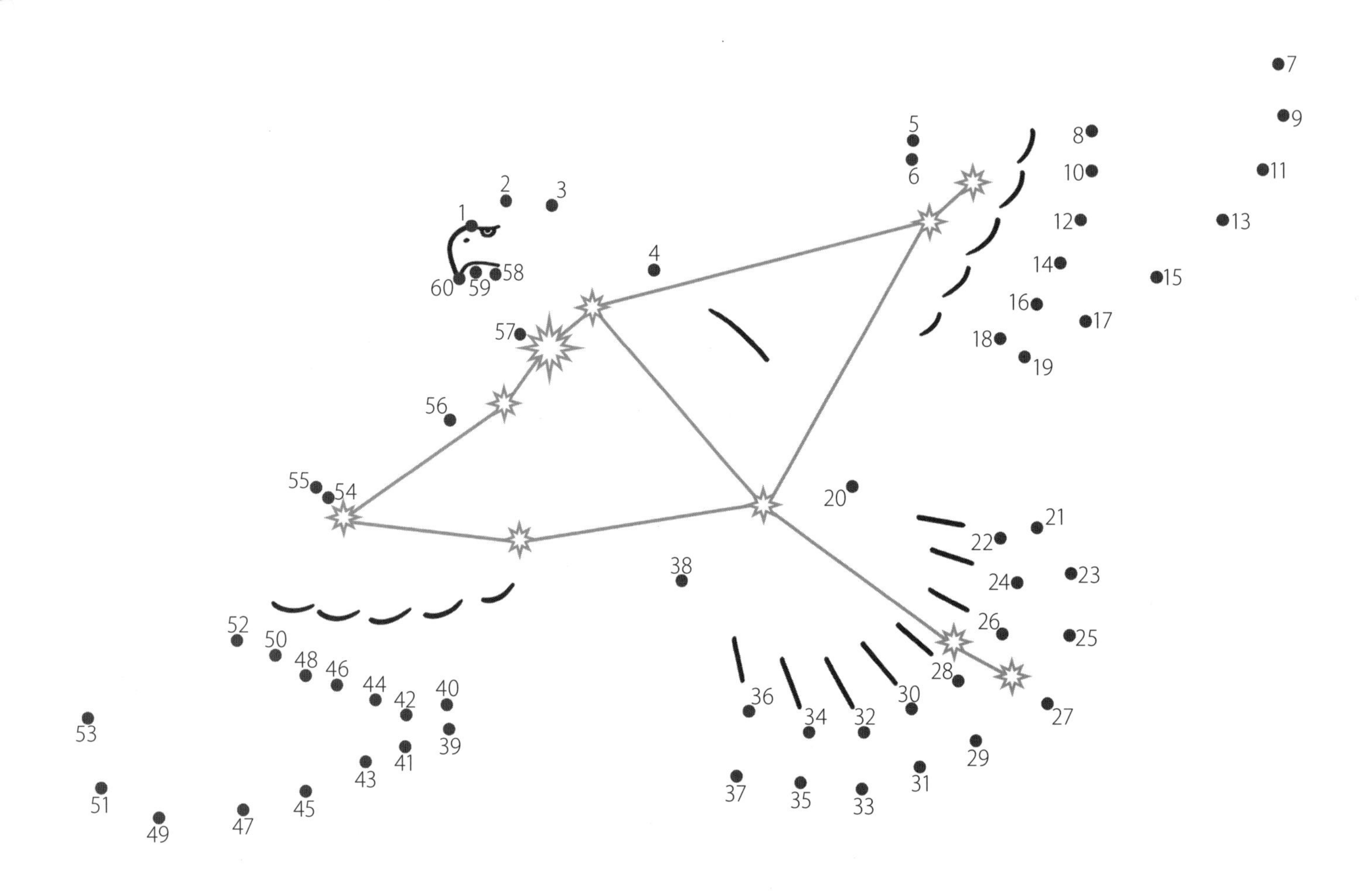

Best Way to Find It (about 10 p.m.)

Look this high . . . July to September . . . facing south . . . for this star pattern.

TIP: The star near Aquila's neck is one of the brightest in the night sky.

It's easy to see Aquila's wings and tail in the constellation's star pattern.

As the Story Goes

The earth was once ruled by giants, called Titans. The god Zeus wanted to free the earth from the Titans, so he and the other gods prepared for war. An eagle named Aquila (*ACK-will-uh*) flew to Zeus. Zeus took it as a sign that he would win the war. The gods did win, and Zeus kept Aquila as his loyal pet.

Man's Best Friends

Color the animals below. Circle the ones that would make good pets.

HORSE

BISON

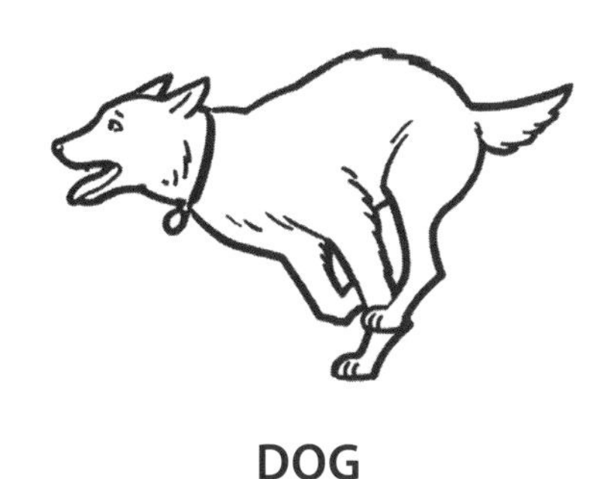

DOG

SCORPION

RATTLESNAKE

BUTTERFLY

RABBIT

HAMSTER

BAT

(answers on page 58)

CYGNUS

THE SWAN

Best Way to Find It (about 10 p.m.)

Look this high . . . August to September . . . facing north . . . for this star pattern.

TIP: Look for the bright star of Cygnus's tail. In August and September, it's almost straight above you.

As the Story Goes

The god Zeus fell in love with a beautiful queen named Leda. But Leda did not love Zeus. To get close to her, Zeus turned into a swan. His trick worked. Leda cared for the swan, not knowing it was Zeus. Eventually, Leda and Zeus had three children together. Zeus put Cygnus (*SIG-nus*) the swan into the sky to honor all swans.

ALSO CHECK OUT: Gemini (page 54)

Beautiful Daughter

One of Zeus and Leda's children was Helen of Troy. She was so beautiful that two armies went to war over her. Draw and color Helen of Troy's beautiful face.

CAPRICORNUS

THE GOAT

Best Way to Find It (about 10 p.m.)

Look this high . . . August to October . . . facing south . . . for this star pattern.

TIP: Capricornus can be hard to find, but Aquila's left wing points straight to it.

Capricornus's dim stars form a triangle, of sorts.

As the Story Goes

The gods' home, Olympus, was attacked by the monster Typhon. Pan could have escaped, but he was loyal. He sent a warning to Zeus instead. Typhon almost caught Pan, but Pan turned half of himself into a goat. He jumped into a river and turned the rest of his body into a fish. Capricornus (*CAP-rih-COR-nus*) is Pan's animal creation.

ALSO CHECK OUT: Pisces (page 40), What's Your Sign? (page 9)

The Names of the Gods

Use the secret code below to learn the names of ten Greek gods.

__ __ __ __ __ __ __ __ __, goddess of __ __ __ __
1 16 8 18 15 4 9 20 5 — 12 15 22 5

__ __ __ __ __ __, god of the __ __ __
1 16 15 12 12 15 — 19 21 14

__ __ __ __, god of __ __ __
1 18 5 19 — 23 1 18

__ __ __ __ __ __ __, goddess of the __ __ __ __
1 18 20 5 13 9 19 — 8 21 14 20

__ __ __ __ __ __, goddess of __ __ __ __ __ __
1 20 8 5 14 1 — 23 9 19 4 15 13

__ __ __ __ __ __ __, goddess of __ __ __ __ __ __ __
4 5 13 5 20 5 18 — 6 1 18 13 9 14 7

__ __ __ __ __, god of the __ __ __ __
8 1 4 5 19 — 4 5 1 4

__ __ __ __ __ __, goddess of __ __ __ __ __ __ __ __
8 5 19 20 9 1 — 2 21 9 12 4 9 14 7

__ __ __ __ __ __, the __ __ __ __ __ __ __ __ __ god
8 5 18 13 5 19 — 13 5 19 19 5 14 7 5 18

__ __ __ __ __ __ __ __, god of the __ __ __
16 15 19 5 9 4 15 14 — 19 5 1

A	B	C	D	E	F	G	H	I	J	K	L	M	N	O	P	Q	R	S	T	U	V	W	X	Y	Z
1	2	3	4	5	6	7	8	9	10	11	12	13	14	15	16	17	18	19	20	21	22	23	24	25	26

(answers on page 58)

AQUARIUS

THE WATER BEARER

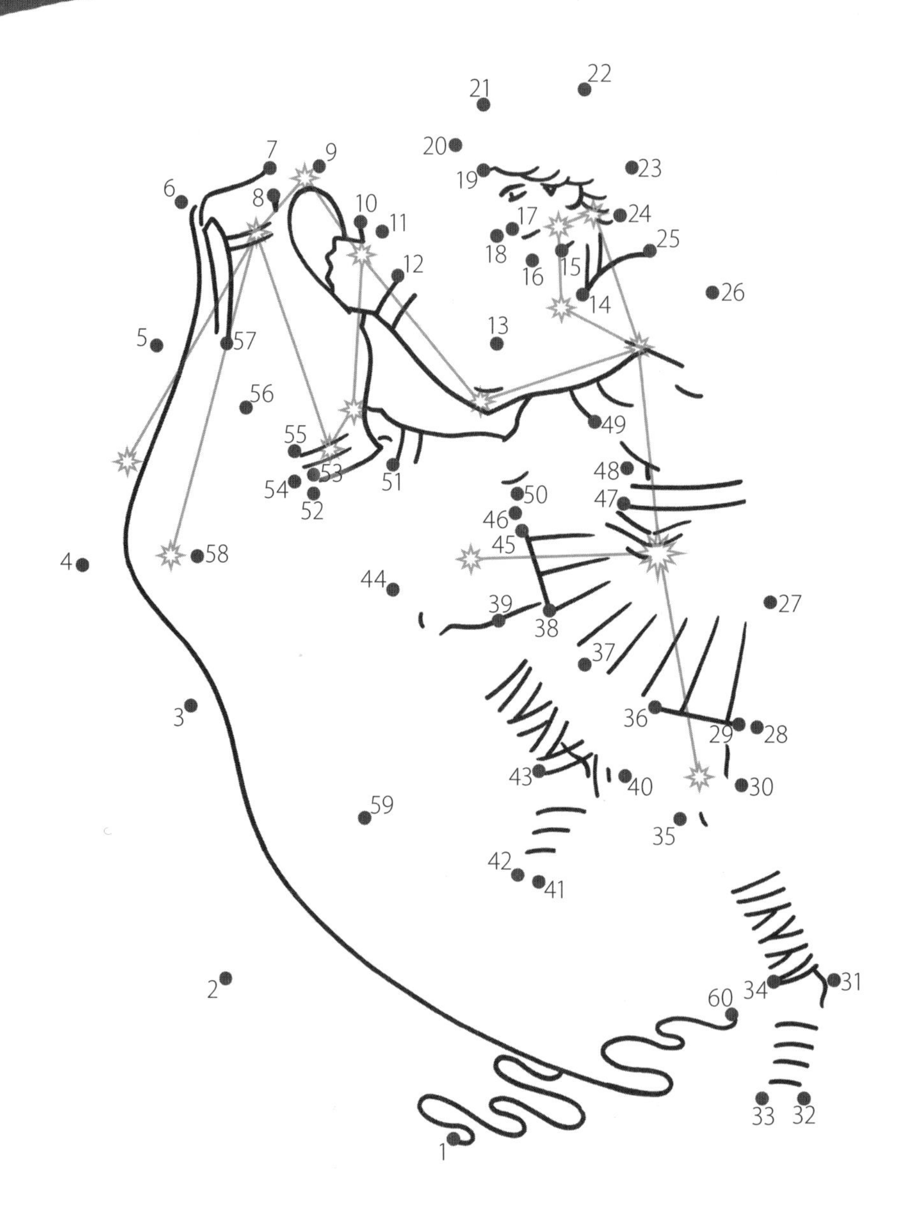

Best Way to Find It (about 10 p.m.)

Look this high . . . September to October . . . facing south . . . for this star pattern.

TIP: You will find Aquarius near the top of Pegasus's head.

As the Story Goes

When Ganymede was born, he was a beautiful baby. By the time he was a young boy, he was known as the most beautiful child in the world. Zeus brought Ganymede to his home on Mount Olympus and allowed him to serve food and drinks to the gods. For Ganymede's service, Zeus placed him in the stars as Aquarius (*uh-KWARE-ee-us*).

ALSO CHECK OUT: What's Your Sign? (page 9)

Fast Food

Zeus is hungry. Help Ganymede bring food, through the maze, to the Greek god.

(answers on page 59)

PEGASUS

THE WINGED HORSE

Best Way to Find It (about 10 p.m.)

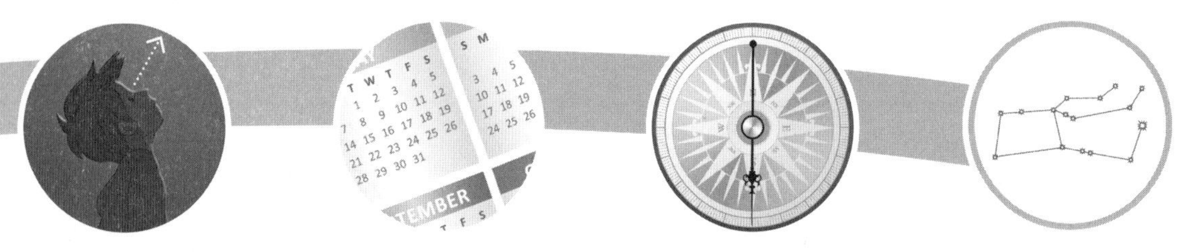

Look this high . . . September to October . . . facing south . . . for this star pattern.

TIP: Look for Pegasus's big square. There aren't many other stars near it.

As the Story Goes

Pegasus (*PEG-uh-sus*) was a winged horse. He was created from the blood of a monster named Medusa. Pegasus sprang to life after the hero Perseus defeated Medusa. After Pegasus learned to fly, he soared to Mount Olympus to live with the gods. Pegasus served Zeus by carrying his lightning bolts across the sky.

ALSO CHECK OUT: Andromeda (page 38), Cassiopeia (page 42), Perseus (page 46)

Animal Mix-up

Pegasus was a mix between a horse and a bird. In the picture below, circle all of the other animals that are mixed up. How many can you find?

(answers on page 59)

ANDROMEDA

THE PRINCESS OF JOPPA

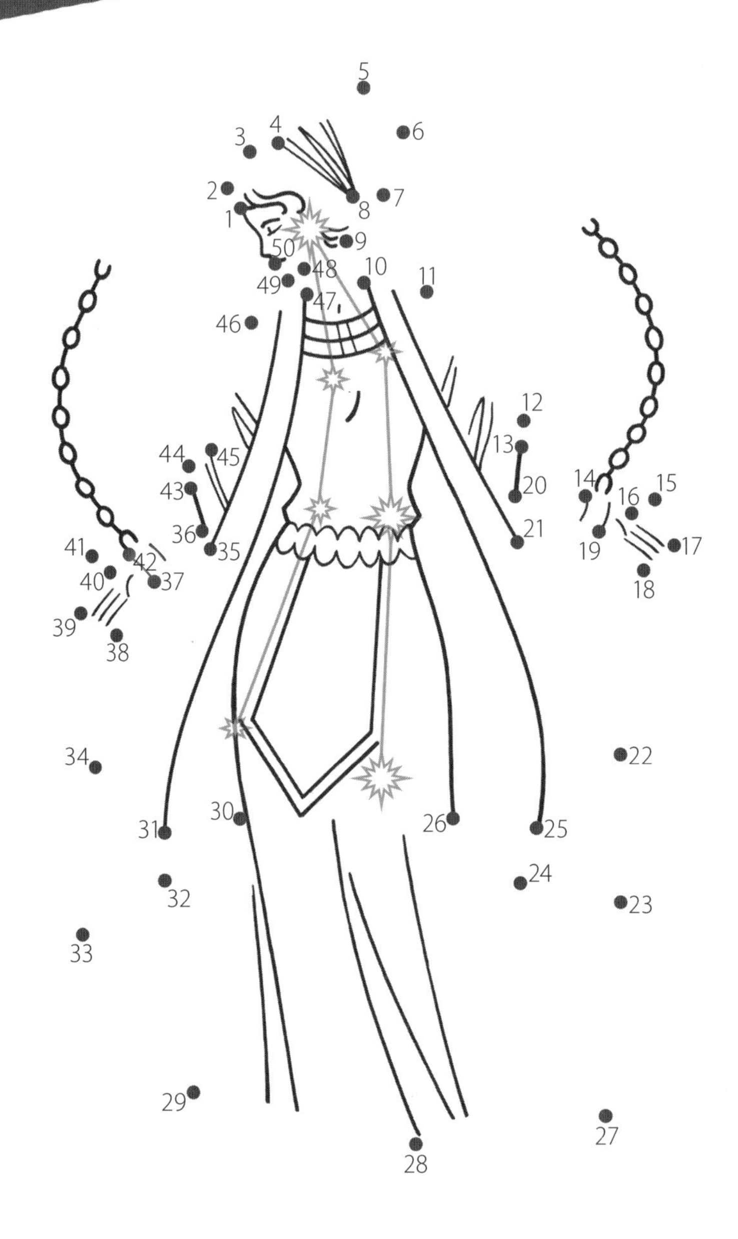

Best Way to Find It (about 10 p.m.)

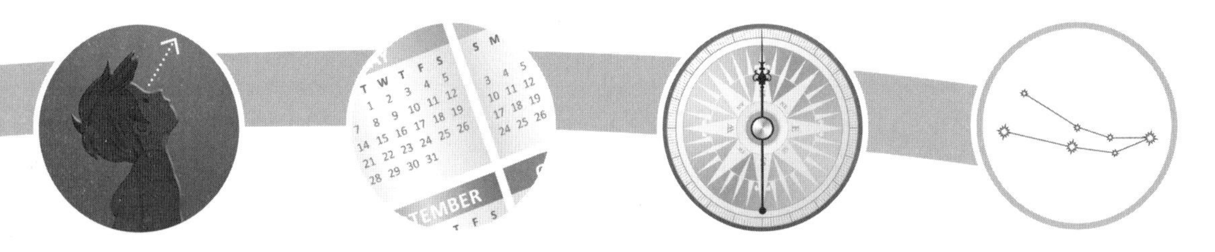

Look this high . . . October to November . . . facing north . . . for this star pattern.

TIP: The star at the tip of Andromeda is also a corner of Pegasus's square.

The stars of Andromeda form a sideways letter "A."

As the Story Goes

A sea monster attacked the land of Joppa. The only way to stop it was to give the monster Andromeda (*an-DROM-ed-uh*), a beautiful princess. Andromeda's parents chained her to a rock, where she waited for the sea monster to take her. But a hero named Perseus defeated the sea monster, rescued Andromeda and married her.

ALSO CHECK OUT: Cassiopeia (page 42), Pegasus (page 36), Perseus (page 46)

What's in a Name?

How many words can you make using the letters in Andromeda's name?

A N D R O M E D A

(answers on page 59)

PISCES

THE FISHES

Best Way to Find It (about 10 p.m.)

Look this high . . . October to November . . . facing south . . . for this star pattern.

TIP: This large constellation curves around the tail end of Pegasus.

As the Story Goes

Typhon was a monster with the heads of 100 giant snakes. He was more powerful than the Greek gods. Typhon attacked the gods on Mount Olympus, and most of the gods ran away. Aphrodite and her son Eros turned themselves into fish. They swam away from Typhon. The constellation Pisces (*PIE-seez*) shows Aphrodite and Eros as fish.

ALSO CHECK OUT: Capricornus (page 32), What's Your Sign? (page 9)

Great Escape

If you could turn into an animal, how would you escape from Typhon? Choose an animal and draw it in the space below. You can use one of the pictured animals or create your own.

CASSIOPEIA

THE QUEEN OF JOPPA

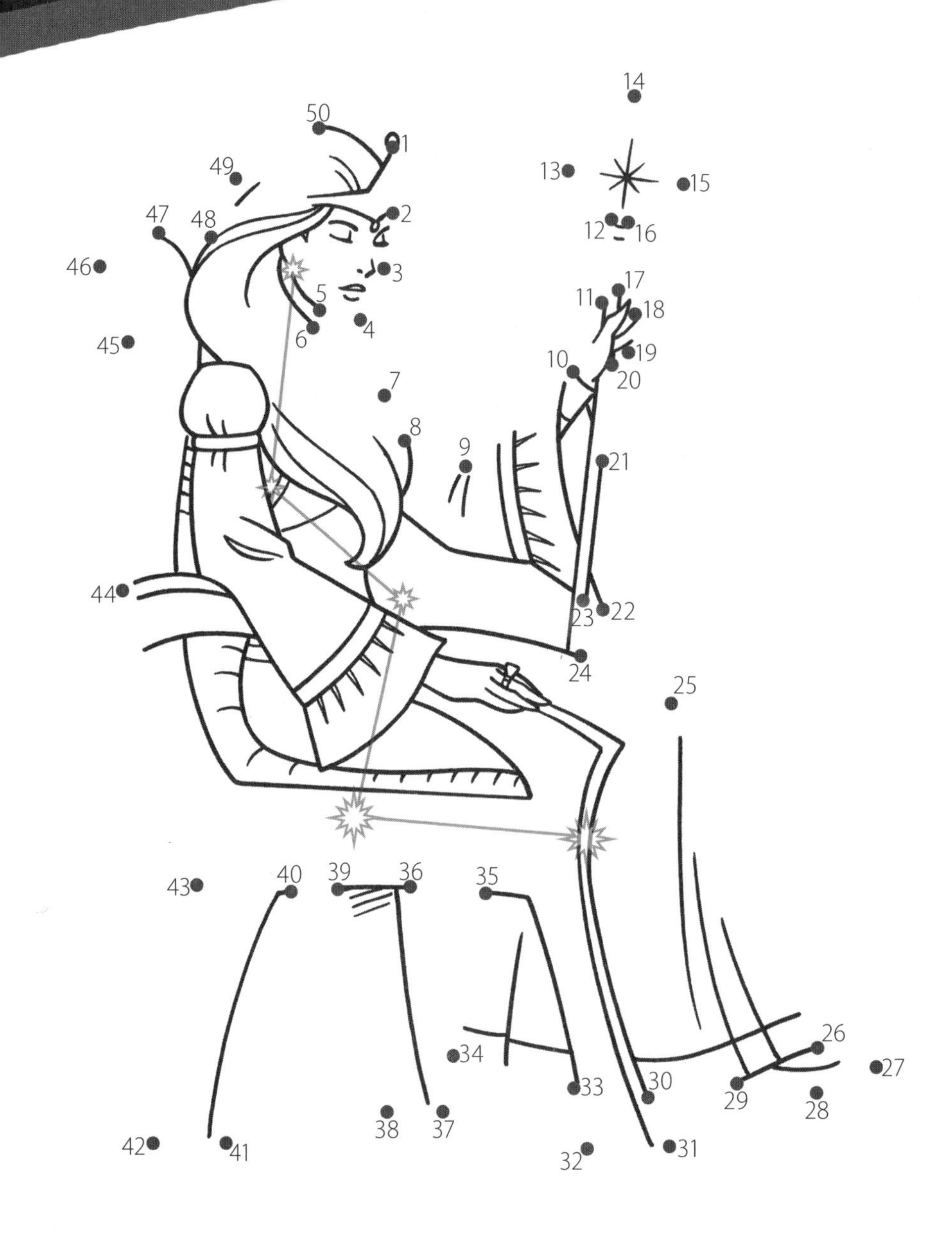

Best Way to Find It (about 10 p.m.)

Look this high . . . October to December . . . facing north . . . for this star pattern.

TIP: Cassiopeia's stars form a squished "M" or "W."

Cassiopeia is a fun constellation to find because the stars are so bright.

As the Story Goes

Cassiopeia (*CASS-ee-oh-PEE-uh*) was a queen. She said her daughter was more beautiful than a goddess. Her bragging made the gods angry. They sent Cetus, a sea monster, to destroy her kingdom. To save the kingdom, Cassiopeia agreed to give her daughter to the sea monster. But a hero named Perseus rescued the princess and defeated Cetus.

ALSO CHECK OUT: Andromeda (page 38), Pegasus (page 36), Perseus (page 46)

Sea Monster?

The story says Cetus was a sea monster. But the word *cetus* means something else in the Latin language. Use the key to color the areas below, and learn what Cetus might have been.

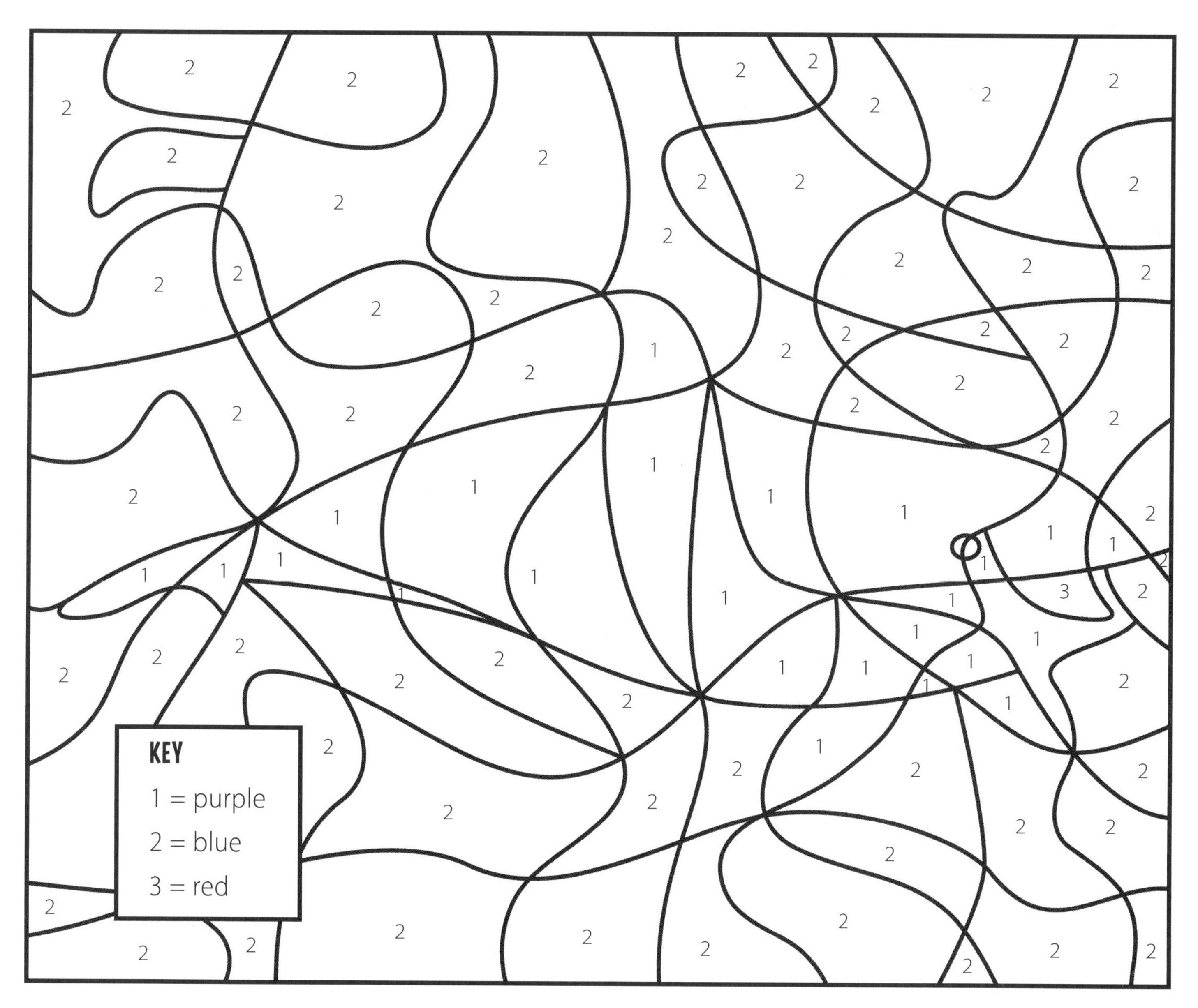

(answers on page 59)

ARIES

THE RAM

Best Way to Find It (about 10 p.m.)

Look this high . . . November to December . . . facing south . . . for this star pattern.

TIP: In early November, Aries will be almost straight up and a bit to your left.

As the Story Goes

Phrixus and Helle were King Athamas' children. Their stepmother, Ino, was cruel and wanted to harm them. Zeus sent a magical ram named Aries (*AIR-eez*) to save the two children. Aries had golden fleece, or fur, and he could fly. Aries rescued the children and flew them away.

ALSO CHECK OUT: What's Your Sign? (page 9)

Zodiac Crossword

Use the clues below to solve the crossword puzzle about the 12 zodiac constellations.

ACROSS

1. Sagittarius was one
5. Hero who defeated Leo
6. Maiden of the night sky
7. One of the twins of Gemini
10. Zeus in disguise as a bull
11. He defeated Orion with his tail

DOWN

1. This constellation is half goat, half fish
2. Pisces fled from this monster
3. On Aries, this was golden
4. This servant was known to the gods as Ganymede
8. Libra represents this
9. She sent Cancer to distract a hero

(answers on page 59)

PERSEUS

THE HERO

Best Way to Find It (about 10 p.m.)

Look this high . . . November to December . . . facing north . . . for this star pattern.

TIP: Look almost straight up. Perseus will be slightly to your left or your right.

Most of the stars in Perseus are bright and easy to see.

As the Story Goes

Perseus (*PER-see-us*) was the son of Zeus and a great hero. In order to save his mother, he had to battle Medusa, a monster with snakes for hair. Medusa could turn people to stone just by looking into their eyes. Perseus defeated her by using a shield as a mirror. Perseus later rescued Princess Andromeda from a sea monster and married her.

ALSO CHECK OUT: Andromeda (page 38), Cassiopeia (page 42), Pegasus (page 36)

The Face of Medusa

Medusa was a scary monster. Her face was covered with scales, and she had snakes for hair. Spread out your fingers (as shown in the picture below), and trace your hand in the space provided. Draw Medusa's face in the palm, and turn your fingers into Medusa's snake hair.

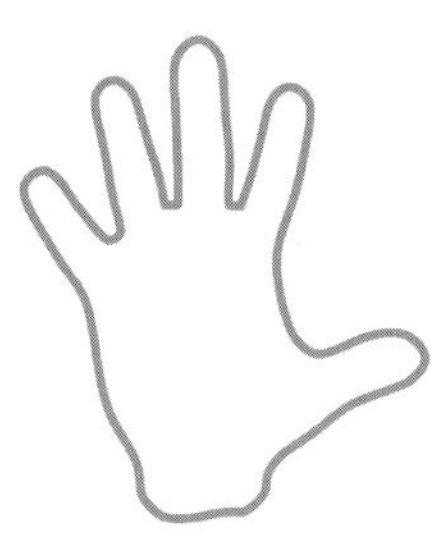

TAURUS

THE BULL

Best Way to Find It (about 10 p.m.)

Look this high . . . December to January . . . facing south . . . for this star pattern.

TIP: Orion's raised right arm is almost touching the tip of Taurus's right horn.

The star below Taurus's right horn is one of the brightest in the sky.

ZODIAC

As the Story Goes

The god Zeus fell in love with Europa. But Europa's father would not allow her to get married. Zeus turned into a bull and hid among her father's cows. Europa rode the bull, not knowing it was Zeus. The god ran to the sea, swam Europa to the island of Crete and made her a queen. The constellation Taurus (*TOR-us*) shows Zeus as a bull.

ALSO CHECK OUT: Canis Major (page 52), What's Your Sign? (page 9)

Animals in the Stars

Taurus is one of many animals that appear in the stars. In the word find below, circle these other animals that form constellations.

F	F	E	A	G	O	R	A	M	A	R	Z	I	P
E	I	K	D	O	T	A	R	E	F	C	O	R	I
O	R	S	W	A	A	B	L	I	F	Z	R	S	O
R	A	D	H	A	B	E	I	Z	L	G	E	A	T
S	N	O	S	K	E	T	Z	A	P	I	N	S	B
W	A	P	S	L	M	O	A	G	H	R	I	O	N
A	H	H	C	B	E	A	R	Z	H	A	R	E	M
D	K	O	O	P	A	H	D	O	L	F	S	A	K
O	D	G	R	U	G	I	S	A	W	F	N	T	U
V	O	J	P	S	L	N	K	W	N	E	F	G	L
A	G	Q	I	K	E	Y	N	J	A	S	C	E	R
R	E	U	O	L	P	N	L	O	S	N	A	K	E
A	M	T	N	O	O	J	D	O	L	P	H	I	N
L	I	O	N	G	R	S	C	O	R	S	E	A	G

BEAR	FISH	LIZARD
CRAB	GIRAFFE	RAM
DOG	HARE	SCORPION
EAGLE	HORSE	SNAKE
DOLPHIN	LION	SWAN

(answers on page 59)

ORION

THE HUNTER

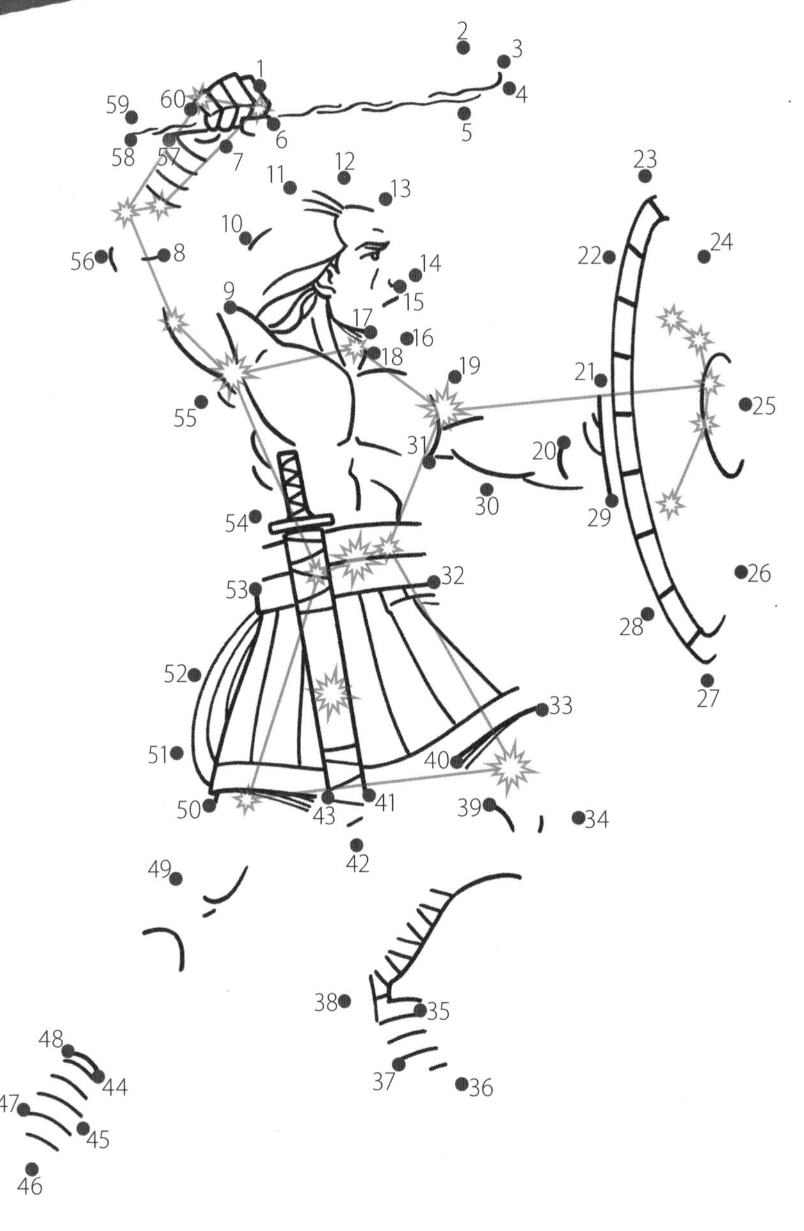

Best Way to Find It (about 10 p.m.)

Look this high . . . December to February . . . facing south . . . for this star pattern.

TIP: Look for Orion's belt, three bright stars all in a row.

As the Story Goes

Orion (*oh-RYE-un*) was a great hunter. He said that he could defeat any animal. The goddess Hera accepted Orion's challenge. She sent a tiny scorpion named Scorpius to attack him. The scorpion stung Orion. Before he died, Orion crushed the scorpion with his club. The gods put Orion and Scorpius in the stars.

ALSO CHECK OUT: Scorpius (page 22)

Hunt for Animals

Help Orion find the animals that are hiding in the picture below.

(answers on page 60)

CANIS MAJOR

THE BIG DOG

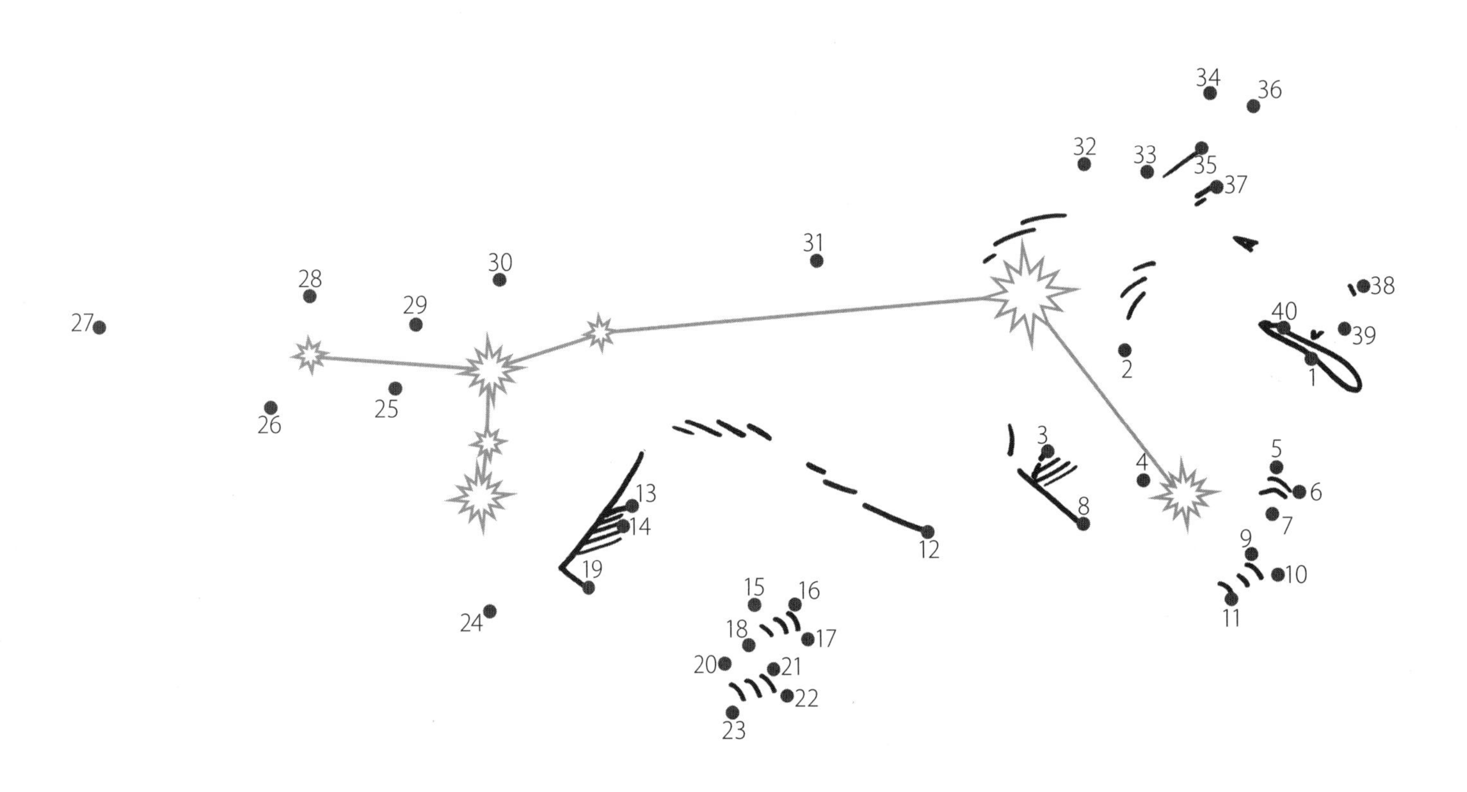

Best Way to Find It (about 10 p.m.)

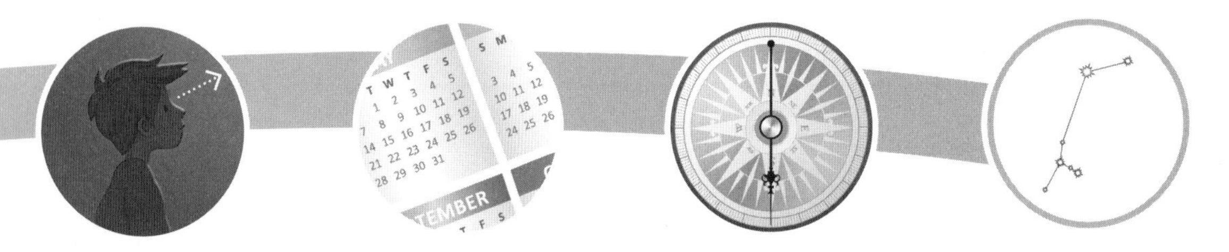

Look this high . . . January to February . . . facing south . . . for this star pattern.

TIP: Draw a line in the sky that passes through Orion's belt. It will lead you to Canis Major.

The star at Canis Major's neck is called Sirius, the brightest of the night sky.

As the Story Goes

The god Zeus fell in love with Europa. He brought her to the island of Crete and made her a queen. Zeus gave her Canis Major (*KANE-iss MAY-jur*), a magical hunting dog that could catch any prey. The gods later created a magic fox that could never be caught. When Canis Major tried to hunt the fox, Zeus placed Canis Major in the stars.

ALSO CHECK OUT: Taurus (page 48)

The Little Dog

Canis Major is not the only dog-themed constellation. Canis Minor can be found in the same corner of the night sky. Use the key to color the areas below, and see the "Little Dog."

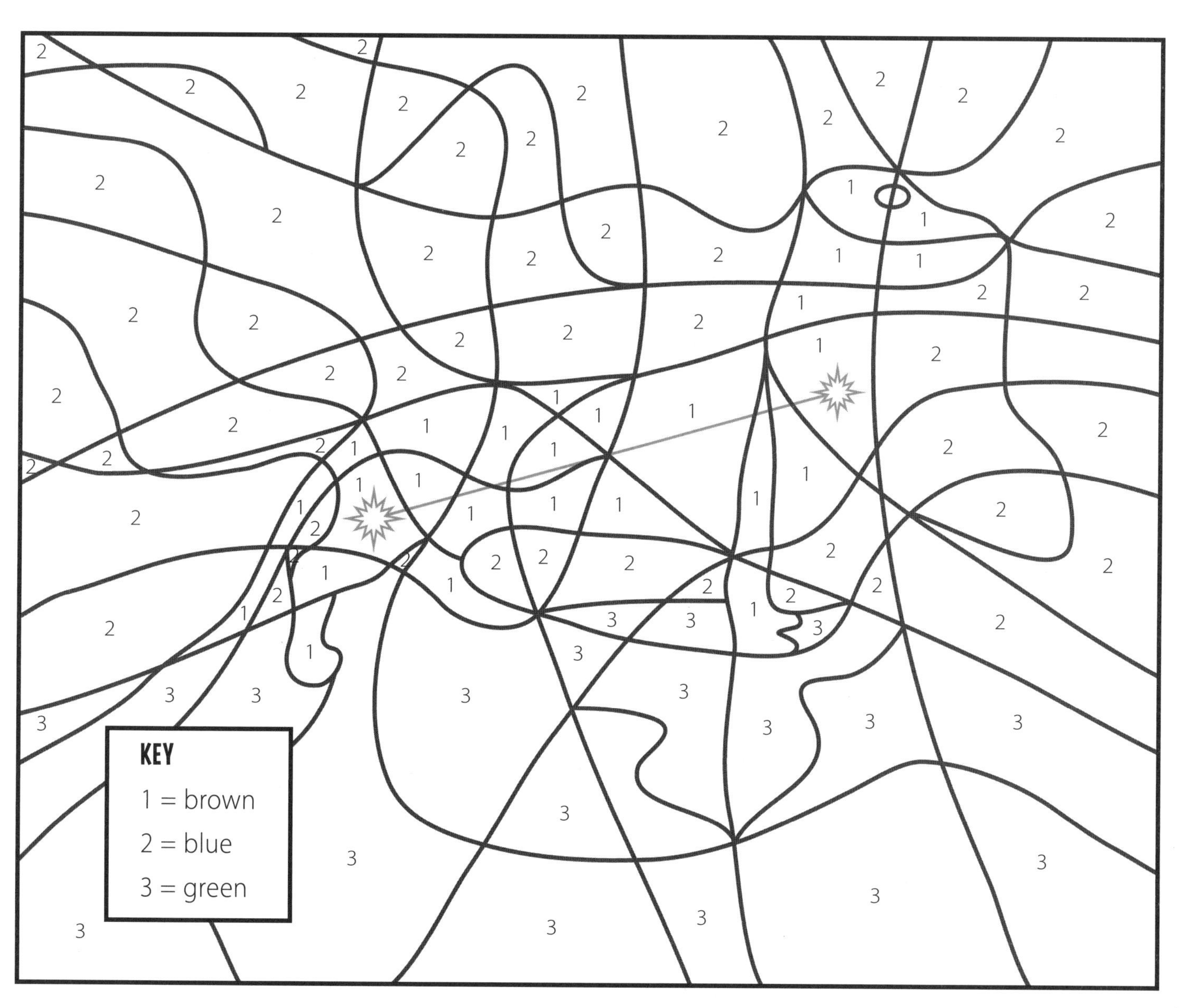

GEMINI

THE TWINS

Best Way to Find It (about 10 p.m.)

Look this high . . . January to February . . . facing south . . . for this star pattern.

TIP: Find Orion's right arm. The twins of Gemini are almost standing on it.

As the Story Goes

Castor and Pollux were twins. Queen Leda was their mother, and Zeus was their father. Castor was born without any powers, but Pollux could not be killed. The twins went on a quest to find Aries' golden fleece. Castor died, but Pollux gave half of his power to Castor and brought him back to life. Gemini (*JEM-in-eye*) is the Latin word for twins.

ALSO CHECK OUT: Aries (page 44), Cygnus (page 30), What's Your Sign? (page 9)

Twin Pictures

Look at each pair of pictures below. Some of them are exactly the same. Others are not. Circle the ones that are different from each other.

(answers on page 60)

CANCER

THE CRAB

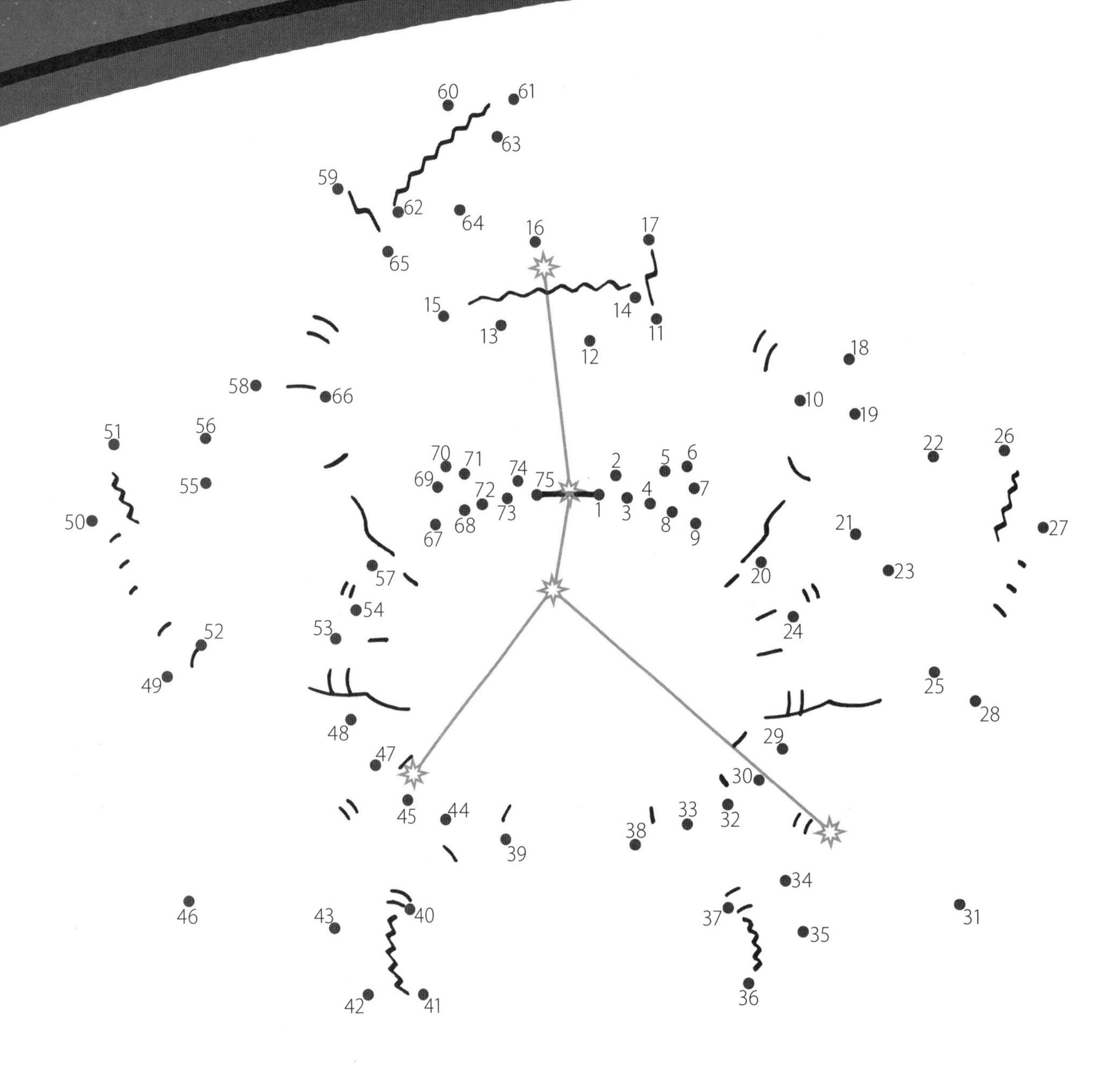

Best Way to Find It (about 10 p.m.)

Look this high . . . February to March . . . facing south . . . for this star pattern.

TIP: Draw a line in the sky that passes through the heads of the Gemini twins. It will lead you to Cancer.

Cancer is one of the dimmest constellations in the night sky.

As the Story Goes

Hercules was a great hero. He battled the Hydra, a monster with many heads. But the goddess Hera did not want Hercules to win. She sent Cancer (*CAN-sir*) the crab to the battle. Cancer pinched Hercules' heels to distract him. Hercules stepped on Cancer and crushed the crab. The hero then defeated the Hydra.

ALSO CHECK OUT: Draco (page 20), Hercules (page 24), Hydra (page 12), Leo (page 8), What's Your Sign? (page 9)

Constellation Connection

Draw a line that connects each star pattern to the picture it represents.

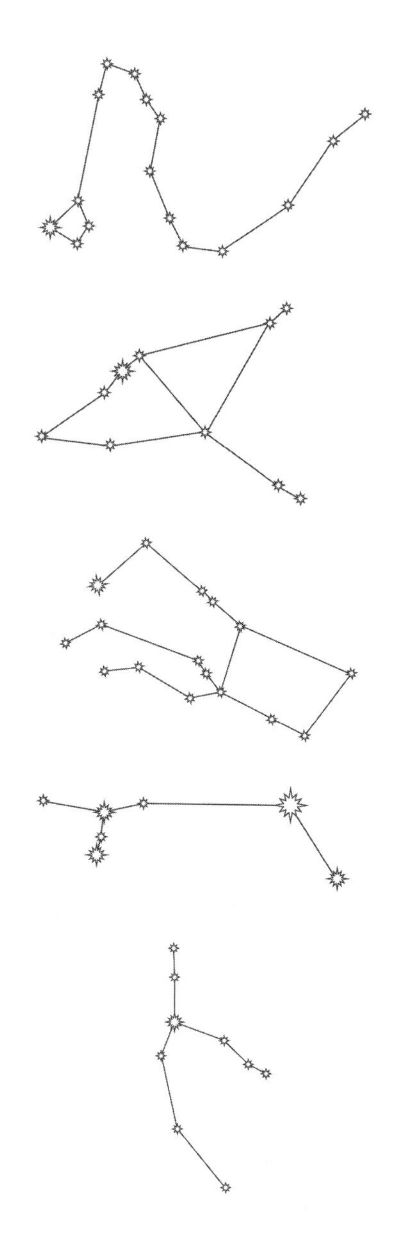

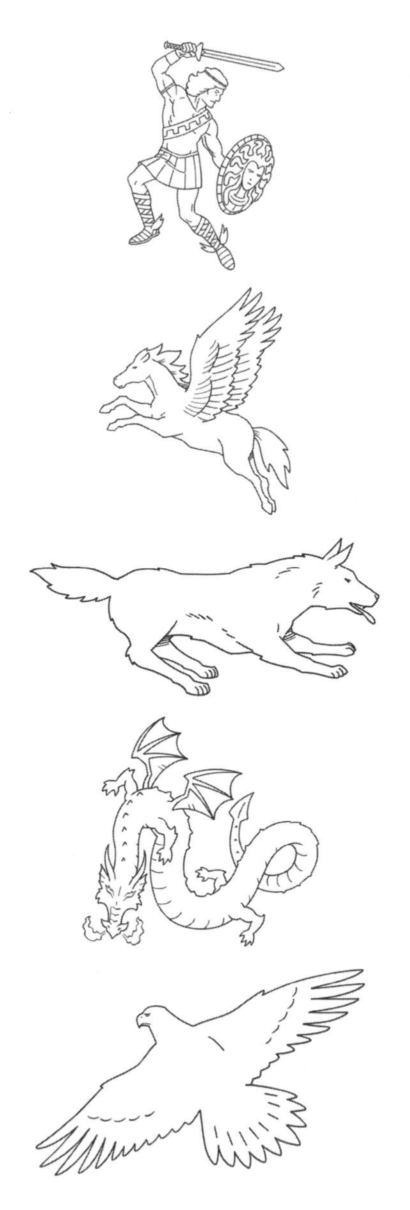

(answers on page 60)

ANSWERS

Name the Constellations (page 7)

1. Ursa Major; 2. Orion; 3. Scorpius; 4. Andromeda; 5. Leo; 6. Sagittarius

Escape from Arcas (page 11)

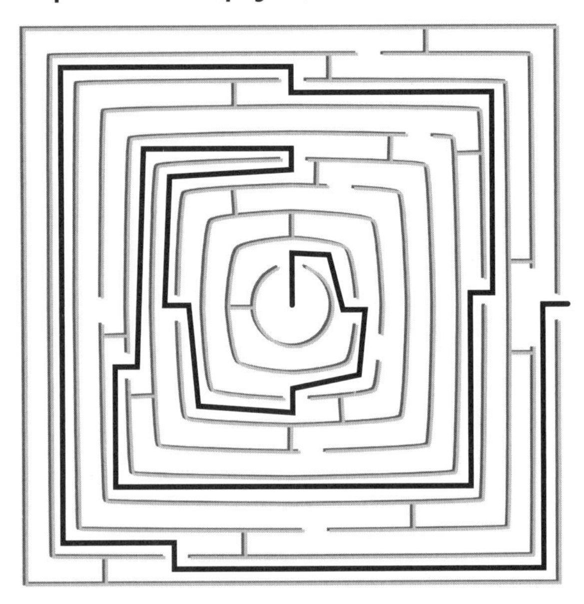

Many Maidens (page 15)

What Did Boötes Invent? (page 17)

Barn, pig, apples, tractor, chicken, windmill. He invented the plow.

Which Weighs More? (page 19)

1. cruise ship; 2. horse; 3. elephant; 4. pie; 5. tractor

Amazing Animal Tails (page 23)

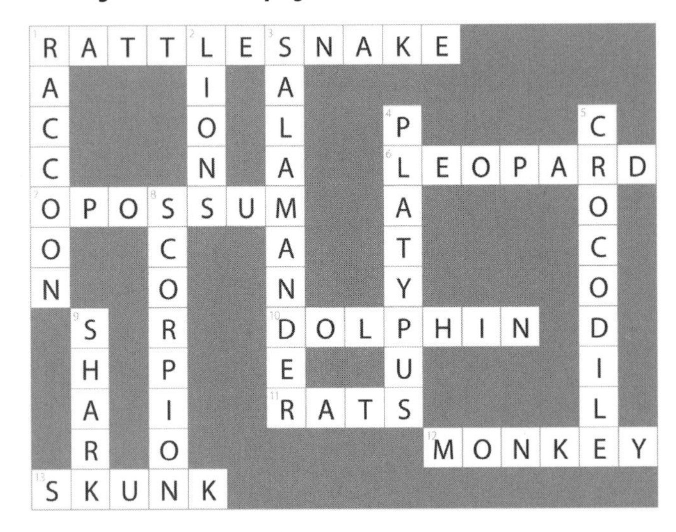

The Twelve Labors of Hercules (page 25)

1. lion; 2. Hydra; 3. deer; 4. pig; 5. cattle; 6. birds; 7. bull; 8. horses; 9. belt; 10. monster; 11. apples; 12. dog

Man's Best Friends (page 29)

horse; dog; butterfly; rabbit; hamster

The Names of the Gods (page 33)

Aphrodite, goddess of love
Apollo, god of the sun
Aries, god of war
Artemis, goddess of the hunt
Athena, goddess of wisdom
Demeter, goddess of farming
Hades, god of the dead
Hestia, goddess of building
Hermes, the messenger god
Poseidon, god of the sea

Fast Food (page 35)

Sea Monster? (page 43)

Cetus might have been a whale.

Zodiac Crossword (page 45)

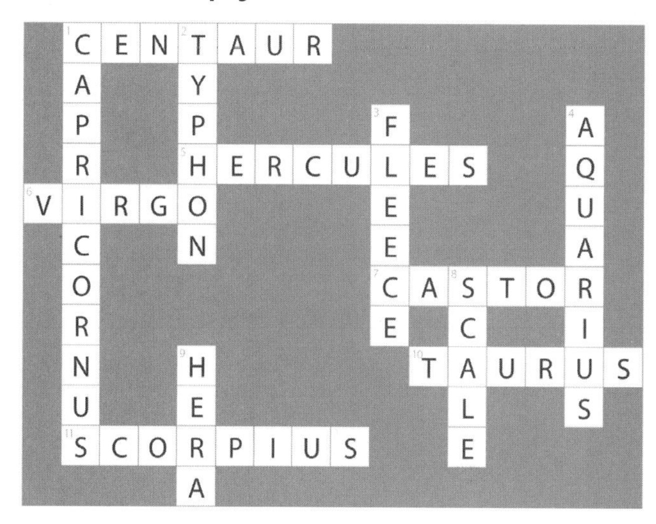

Animal Mix-up (page 37)

What's in a Name? (page 39)

a, ad, add, adore, adored, am, amen, amend, an, and, are, arm, armed, dad, dame, dare, dared, dead, dear, demand, demon, do, doe, dome, domed, done, dorm, dream, ear, earn, end, era, ma, mad, made, man, manor, me, mean, men, mend, moaned, modern, more, name, named, near, no, nod, oar, odd, odder, on, one, or, ran, random, read, red, road, rod, roam, roamed, rode (and many more)

Animals in the Stars (page 49)

ANSWERS

Hunt for Animals (page 51)

Twin Pictures (page 55)

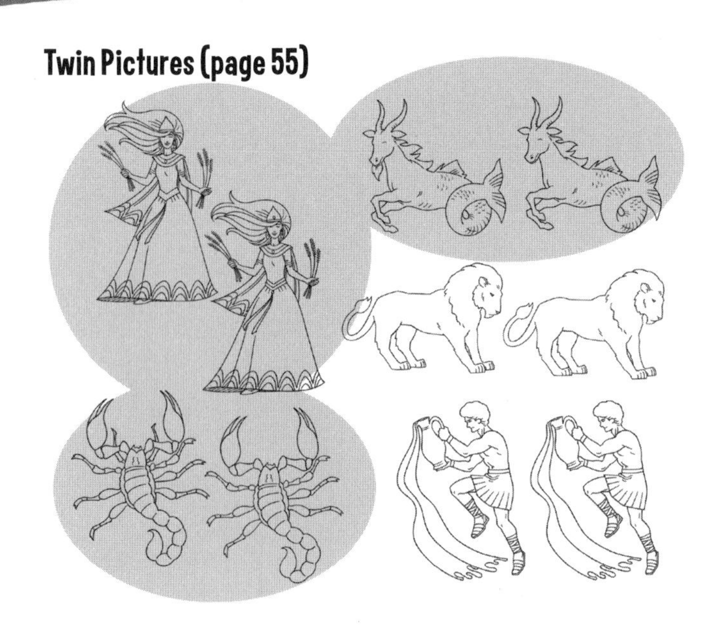

Constellation Connection (page 57)

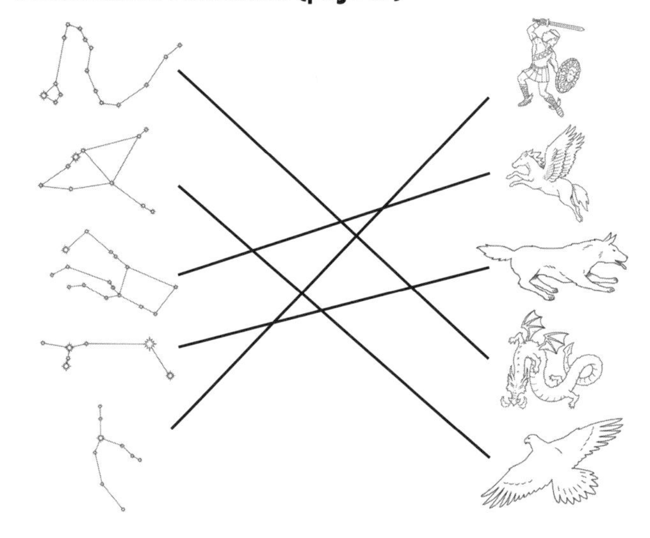

Flash Cards

The following pages feature flash cards of the 26 constellations in this book. They can be cut out along the dotted lines. Put the cards in a pile and look at each, one at a time. Say the name of each constellation, and then flip the card over to find out if you are right.

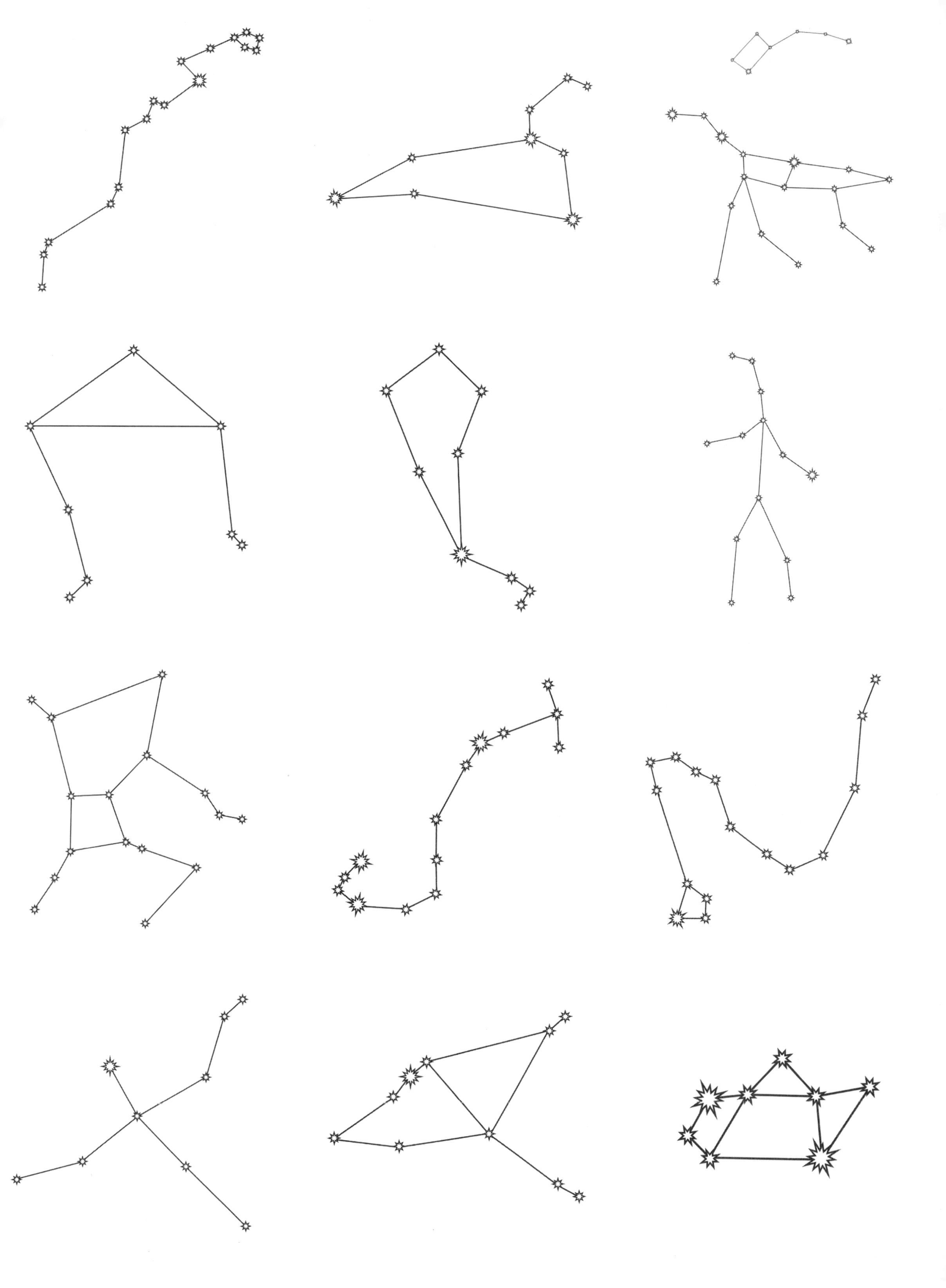

URSA MAJOR + URSA MINOR	LEO	HYDRA
VIRGO	BOÖTES	LIBRA
DRACO	SCORPIUS	HERCULES
SAGITTARIUS	AQUILA	CYGNUS

CAPRICORNUS	AQUARIUS	PEGASUS + ANDROMEDA
PISCES	CASSIOPEIA	ARIES
PERSEUS	TAURUS	ORION
CANIS MAJOR	GEMINI	CANCER

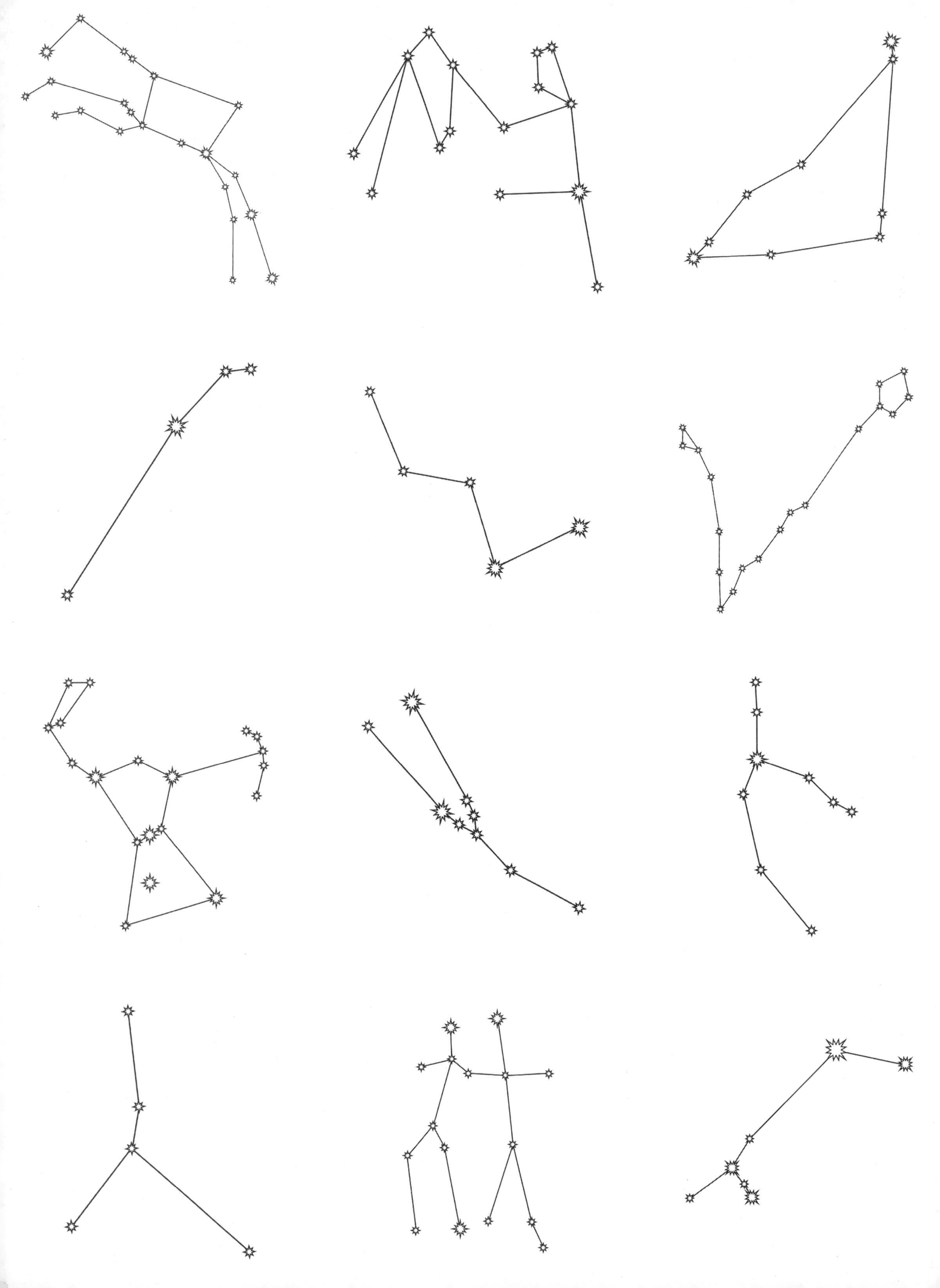